Statistics: Unlocking the Power of Data

Student Solutions Manual

Robin H. Lock
St. Lawrence University

Patti Frazer Lock
St. Lawrence University

Kari Lock Morgan
Duke University

Eric F. Lock
University of North Carolina

Dennis F. Lock
Iowa State University

WILEY

PUBLISHER	Laurie Rosatone
ACQUISITIONS EDITOR	Joanna Dingle
SENIOR DEVELOPMENTAL EDITOR	Mary O'Sullivan
FREELANCE DEVELOPMENTAL EDITOR	Anne Scanlan-Rohrer
PROJECT EDITOR	Shannon Corliss
ASSOCIATE CONTENT EDITOR	Beth Pearson
EDITORIAL ASSISTANT	Elizabeth Baird
SENIOR CONTENT MANAGER	Karoline Luciano
SENIOR PRODUCTION EDITOR	Kerry Weinstein
MARKETING MANAGER	Melanie Kurkjian
SENIOR PRODUCT DESIGNER	Tom Kulesa
OPERATIONS MANAGER	Melissa Edwards

Founded in 1807, John Wiley & Sons, Inc. has been a valued source of knowledge and understanding for more than 200 years, helping people around the world meet their needs and fulfill their aspirations. Our company is built on a foundation of principles that include responsibility to the communities we serve and where we live and work. In 2008, we launched a Corporate Citizenship Initiative, a global effort to address the environmental, social, economic, and ethical challenges we face in our business. Among the issues we are addressing are carbon impact, paper specifications and procurement, ethical conduct within our business and among our vendors, and community and charitable support. For more information, please visit our website: www.wiley.com/go/citizenship.

ISBN 978-0-470-63318-2

10 9 8 7 6 5 4 3 2 1

Contents

Section 1.1 Solutions

1.1 (a) The cases are the people who are asked the question.

(b) The variable is whether each person supports the law or not. It is categorical.

1.3 (a) The cases are the teenagers in the sample.

(b) The variable is the result (yes or no) indicating whether each teenager eats at least five servings a day of fruits and vegetables. It is categorical.

1.5 (a) The 10 beams that were tested.

(b) The force at which each beam broke. It is quantitative.

1.7 Since we expect the number of years smoking cigarettes to impact lung capacity, we think of the number of years smoking as the explanatory variable and the lung capacity as the response variable.

1.9 Ingesting more alcoholic drinks will cause the level of alcohol in the blood to increase, so the number of drinks is the explanatory variable and blood alcohol content is the response.

1.11 (a) *Year* and *HigherSAT* are categorical. The other six variables are all quantitative, although *Siblings* might be classified as either categorical or quantitative.

(b) There are many possible answers, such as "What proportion of the students are first year students?" or "What is the average weight of these students?"

(c) There are many possible answers, such as "Do seniors seem to weigh more than first year students?" or "Do students with high Verbal SAT scores seem to also have high Math SAT scores?"

1.13 (a) The categorical variables are *Smoke*, *Vitamin*, *Gender*, *VitaminUse*, and *PriorSmoke*. The other eleven variables are quantitative.

(b) There are many possible answers. For example, one possible relationship of interest between two categorical variables is the relationship between smoking status and vitamin use. A possibly interesting relationship between two quantitative variables is between the amount of beta-carotene consumed in the diet (*BetaDiet*) and the concentration of beta-carotene in the blood (*BetaPlasma*). One possible relationship of interest between a categorical variable and a quantitative variable might be gender and the number of alcoholic drinks per week.

1.15 The individual cases are the lakes from which water samples were taken. For each lake in the sample, we record the concentration of estrogen in the water and the fertility level of fish. Both are quantitative variables.

1.17 One variable is whether each male was fed a high-fat diet or a normal diet. This is the explanatory variable and it is categorical. The response variable is whether or not the daughters developed metabolic syndrome, which is also categorical.

1.19 In the first study, the cases are the students. The only variable is whether or not the student has smoked a hookah. This is a categorical variable.
In the second study, the cases are the people in a hookah bar. The variables are the length of the session, the frequency of puffing, and the depth of inhalation. All are quantitative.
In the third study, the cases are the smoke samples, and the variables are the amount of tar, nicotine, and heavy metals. All three variables are quantitative.

1.21 If we simply record age in years and income in dollars, the variables are quantitative. Often, however, in a survey, we don't ask for the exact age but rather what age category the participant falls in (20 - 29, 30 - 39, etc). Similarly, we often don't ask for exact income but for an income category (less than $10,000, between $10,000 and $25,000, etc.) If we ask participants what category they are in for each variable, then the variables are categorical.

1.23   (a) The "most appealing" question would require just one categorical variable with four possible categories corresponding to the four flavors.

(b) Data for "which are appealing" would need four categorical variables, one for each flavor, with values of *yes* or *no*.

(c) The "rank the flavors" item would need four variables recording the rank given to each flavor. These could be considered categorical (first, second, ...) or quantitative (numerical value of the rank).

(d) The "rate the flavors" item would need four quantitative variables, each with a value between 1 and 10 for the rating assigned to that flavor.

1.25 We could survey a sample of people and ask their household income and measure happiness in some way, such as asking how happy they are on a scale of 1-10. The cases would be the people we collect data from. The variables in this case would be household income and happiness rating, although any two variables measuring wealth and happiness are possible.

Section 1.2 Solutions

1.27 This is a sample, because only a subset of fish are measured.

1.29 This is a population, because all registered vehicles are accounted for.

1.31 The sample is the 120 people interviewed. The population might be all people in that town or all people that go to the mall in that town or a variety of other groups larger than and containing the 120 people in the sample.

1.33 The sample is the five hundred Canadian adults that were asked the question; the population is all Canadian adults.

1.35 The sample consists of the cookies that were in the package; the population is all Chips Ahoy! cookies.

1.37 (a) The sample is the 100 college students who were asked the question.

(b) The population we are interested in is all Americans.

(c) A population we can generalize to, given our sample, is college students.

1.39 (a) The sample is the 1500 people who were contacted.

(b) The population we are interested in is all residents of the US.

(c) A population we can generalize to, given our sample, is residents of Minnesota.

1.41 Yes, this is a random sample from the population.

1.43 No, this is not a random sample, because some employees may be more likely than others to actually complete the survey.

1.45 No, this is not a random sample. We might think we can pick out a "representative sample", but we probably can't. We need to let a random number generator do it for us.

1.47 This sample is definitely biased because only students who are at the library on a Friday night can be selected. The random sample should be from all students.

1.49 This sample is biased because taking 10 apples off the top is not a random sample. The apples on the bottom of the truckload are probably more likely to be bruised.

1.51 This sample is biased because it is a volunteer survey in which people choose to participate or not. Most likely, the people taking the time to respond to the email will have stronger opinions than the rest of the student body.

1.53 (a) The individual cases are the over 6000 restroom patrons who were observed. The description makes it clear that at least three variables are recorded. One is whether or not the person washed their hands, another is the gender of the individual, and a third is the location of the observation. All three are categorical.

(b) In a phone survey, people are likely to represent themselves in the best light and not always give completely honest answers. That is why it is important to also find other ways of collecting data, such as this method of observing people's actual habits in the restroom.

1.55 No, we cannot conclude that about 79% of all people think physical beauty matters, since this was a volunteer sample in which only people who decided to vote were included in the sample, and only people looking at *cnn.com* even had the opportunity to vote. The sample is the 38,485 people who voted. The population if we made such an incorrect conclusion would be all people. There is potential for sampling bias in every volunteer sample.

1.57 Yes! The sample is a random sample so we can be quite confident that it is probably a representative sample.

1.59 (a) The population in the CPS is all US residents. (Also acceptable: US citizens, US households...)

  (b) The population in the CES survey is all non-farm businesses and government agencies in the U.S.

  (c)   i. The CES survey would be more relevant, because the question pertains to companies.
        ii. The CPS would be more relevant, because the question pertains to American people.
        iii. The CPS would be more relevant, because the question pertains to people, not businesses.

1.61 Answers will vary. See the technology notes to see how to use specific technology to select a random sample.

1.63 (a) Number the rows from 1 to 100 and the plants within each row from 1 to 300. Use a computer random number generator to pick a number between 1 and 100 to select a row and a second number between 1 and 300 to identify the plant within that row. Repeat until 30 different plants have been selected. Other options are possible: for example, we could number the plants from 1 to 3000 and randomly select 30 numbers between 1 and 3000.

  (b) Here is the start of one sample.

```
Row     Plant
#94     #180
#83     # 81
#10     #222
```

Other options are possible.

Section 1.3 Solutions

1.65 Since "no link is found" there is neither association nor causation.

1.67 The phrase "more likely" indicates an association, but there is no claim that wealth *causes* people to lie, cheat or steal.

1.69 The statements imply that eating more fiber will cause people to lose wait, so this is a causal association.

1.71 One possible confounding variable is population. Increasing population in the world over time may mean more beef and more pork is consumed. Other answers are possible. Remember that a confounding variable should be associated with both of the variables of interest.

1.73 One possible confounding variable is snow in the winter. When there is more snow, sales of both toboggans and mittens will be higher. Remember that a confounding variable should be associated with both of the variables of interest.

1.75 One possible confounding variable is gender. Males usually have shorter hair and are taller. Other answers are possible. Remember that a confounding variable should be associated with both of the variables of interest.

1.77 We are actively manipulating the explanatory variable (playing music or not), so this is an experiment.

1.79 We are not manipulating any variables in this study, we are only measuring things (omega-3 oils and water acidity) as they exist. This is an observational study.

1.81 The penguins in this study were randomly assigned to get either a metal or an electronic tag so this is an experiment.

1.83 Snow falls when it is cold out and the heating plant will be used more on cold days than on warm days. Also, when snow falls, people have to shovel the snow and that can lead to back pain. Notice that the confounding variable has an association with *both* the variables of interest.

1.85 Age or grade level! Certainly, students in sixth grade can read substantially better than students in first grade and they are also substantially taller. Grade level is influencing both of the variables and it is a confounding variable. If we look at individual grades one at a time, the association could easily disappear.

1.87 (a) The cases are university students. One variable is whether the student lives in a single-sex or co-ed dorm. This is a categorical variable. The other variable is how often the student reports hooking up for casual sex, which is quantitative.

  (b) The type of dorm is the explanatory variable and the number of hook-ups is the response variable.

  (c) Yes, apparently the studies show that students in same sex dorms hook-up for casual sex more often, so there is an association.

  (d) Yes, the president is assuming that there is a causal relationship, since he states that "single sex dorms *reduce* the number of student hook-ups".

  (e) There is no indication that any variable was manipulated, so the studies are probably observational studies.

  (f) The type of student who requests a single sex dorm might be different from the type of student who requests a co-ed dorm. There are other possible confounding variables.

(g) No! We should not assume causation from an observational study.

(h) He is assuming causation when there may really only be association.

1.89 (a) This is an experiment since the background color was actively assigned by the researchers.

(b) The explanatory variable is the background color, which is categorical. The response variable is the attractiveness rating, which is quantitative.

(c) The men were randomly divided into the two groups. Blinding was used by not telling the participants or those working with them the purpose of the study.

(d) Yes. Since this was a well-designed randomized experiment, we can conclude that there is a causal relationship.

1.91 (a) The explanatory variable is whether or not the person had a good night's sleep or is sleep-deprived. The response variable is attractiveness rating.

(b) Since the explanatory variable was actively manipulated, this is an experiment. The two treatments are well-rested and sleep-deprived. Since all 23 subjects were photographed with both treatments, this is a matched pairs experiment.

(c) Yes, we can conclude that sleep-deprivation causes people to look less attractive, because this is an experiment.

1.93 (a) The explanatory variable is amount of sleep and the response variable is growth in height.

(b) We would take a sample of children and randomly divide them into two groups. One group would get lots of sleep and the other would be deprived of sleep. Then after some time passed, we would compare the amount of height increase for the children in the two groups.

(c) An experiment is necessary in order to verify a cause and effect relationship, but it would definitely not be appropriate to randomly assign some of the kids to be sleep-deprived for long periods of time just for the purposes of the experiment!

1.95 (a) Randomly divide the students into two groups of 20 students each. One group gets alcohol and the other gets water. Measure reaction time for students in both groups.

(b) Measure reaction time for all 40 students both ways: after drinking alcohol and after drinking water. Do the tests on separate days and randomize the order in which the students are given the different treatments. Measure the difference in reaction time for each student.

Section 2.1 Solutions

2.1 The total number is $169 + 193 = 362$, so we have $\hat{p} = 169/362 = 0.4669$. We see that 46.69% are female.

2.3 The total number is $94 + 195 + 35 + 36 = 360$ and the number who are juniors or seniors is $35 + 36 = 71$. We have $\hat{p} = 71/360 = 0.1972$. We see that 19.72% percent of the students who identified their class year are juniors or seniors.

2.5 Since this describes a proportion for all residents of the US, the proportion is for a population and the correct notation is p. We see that the proportion of US residents who are foreign born is $p = 0.124$.

2.7 The report describes the results of a sample, so the correct notation is $\hat{p}$. The proportion of US adults who believe the government does not provide enough support for soldiers returning from Iraq or Afghanistan is $\hat{p} = 931/1502 = 0.62$.

2.9 A relative frequency table is a table showing the proportion in each category. We see that the proportion preferring an Academy award is $31/362 = 0.086$, the proportion preferring a Nobel prize is $149/362 = 0.412$, and the proportion preferring an Olympic gold medal is $182/362 = 0.503$. These are summarized in the relative frequency table below. In this case, the relative frequencies actually add to 1.001 due to round-off error.

Response	Relative Frequency
Academy award	0.086
Nobel prize	0.412
Olympic gold medal	0.503
Total	1.00

2.11 One possible table is shown below. You might also choose to include the totals both down and across. It is also perfectly correct to switch the rows and columns.

	1	2	3
A	3	1	8
B	4	3	1

2.13 (a) The sample is the 119 players who were observed. The population is all people who play rock-paper-scissors. The variable records which of the three options each player plays. This is a categorical variable.

(b) A relative frequency table is shown below. We see that rock is selected much more frequently than the others, and then paper, with scissors selected least often.

Option selected	Relative frequency
Rock	0.555
Paper	0.328
Scissors	0.118
Total	1.0

(c) Since rock is selected most often, your best bet is to play paper.

(d) Your opponent is likely to play paper again, so you should play scissors.

2.15 (a) The variable records whether or not tylosin appears in the dust samples. The individual cases in the study are the 20 dust samples.

(b) Here is a frequency table for the presence or absence of tylosin in the dust samples.

Category	Frequency
Tylosin	16
No tylosin	4
Total	20

(c) A bar chart for the frequencies is shown below.

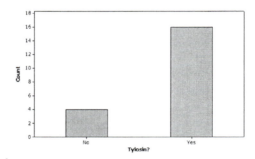

(d) The table below shows the relative frequencies for cases with and without tylosin.

Category	Relative frequency
Tylosin	0.80
No tylosin	0.20
Total	1.00

2.17 (a) There are two variables, both categorical. One is whether or not the dog selected the cancer sample and the other is whether or not the test was a breath test or a stool test.

(b) We need to include all possible outcomes for each variable when we make a two way table. The result variable has two options (dog is correct or dog is not correct) and the type of test variable has two options (breath or stool). The two-way table below summarizes these data.

	Breath test	Stool test	Total
Dog selects cancer	33	37	70
Dog does not select cancer	3	1	4
Total	36	38	74

(c) The dog got $33/36 = 0.917$ or 91.7% of the breath samples correct and $37/38 = 0.974$ or 97.4% of the stool samples correct.

(d) The dog got 70 tests correct and 37 of those were stool samples, so $37/70 = 0.529$ of the tests the dog got correct were stool samples.

2.19 (a) We compute the percentage of smokers in the female column and in the male column. For females, we see that $16/169 = 0.095$, so 9.5% of the females in the sample classify themselves as smokers. For males, we see that $27/193 = 0.140$, so 14% of the males in the sample classify themselves as smokers. In this sample, a larger percentage of males are smokers.

(b) For the entire sample, the proportion of smokers is $43/362 = 0.119$, or 11.9%.

(c) There are 43 smokers in the sample and 16 of them are female, so the proportion of smokers who are female is $16/43 = 0.372$, or 37.2%.

2.21 The two-way table with row and column totals is shown.

	Near-death experience	No such experience	Total
Cardiac arrest	11	105	116
Other cardiac problem	16	1463	1479
Total	27	1568	1595

To compare the two groups, we compute the percent of each group that had a near-death experience. For the cardiac arrest patients, the percent is $11/116 = 0.095 = 9.5\%$. For the patients with other cardiac problems, the percent is $16/1479 = 0.011 = 1.1\%$. We see that approximately 9.5% of the cardiac arrest patients reported a near-death experience, which appears to be much higher than the 1.1% of the other patients reporting this.

2.23 (a) This is an experiment. Participants were actively assigned to receive either electrical stimulation or sham stimulation.

(b) The study appears to be single-blind, since it explicitly states that participants did not know which group they were in. It is not clear from the description whether the study was double-blind.

(c) There are two variables. One is whether or not the participants solved the problem and the other is which treatment (electrical stimulation or sham stimulation) the participants received. Both are categorical.

(d) Since the groups are equally split, there are 20 participants in each group. We know that 20% of the control group solved the problem, and 20% of 20 is $0.20(20) = 4$ so 4 solved the problem and 16 did not. Similarly, in the electrical stimulation group, $0.6(20) = 12$ solved the problem and 8 did not. See the table.

Treatment	Solved	Not solved
Sham	4	16
Electrical	12	8

(e) We see that $4 + 12 = 16$ people correctly solved the problem, and 12 of the 16 were in the electrical stimulation group, so the answer is $12/16 = 0.75$. We see that 75% of the people who correctly solved the problem had the electrical stimulation.

(f) We have $\hat{p}_E = 0.60$ and $\hat{p}_S = 0.20$ so the difference in proportions is $\hat{p}_E - \hat{p}_S = 0.60 - 0.20 = 0.40$.

(g) The proportions who correctly solved the problem are quite different between the two groups, so electrical stimulation does seem to help people gain insight on a new problem type.

2.25 (a) Since no one assigned smoking or not to the participants, this is an observational study. Because this is an observational study, we can not use this data to determine whether smoking influences one's ability to get pregnant. We can only determine whether there is an association between smoking and ability to get pregnant.

(b) The sample collected is on women who went off birth control in order to become pregnant, so the population of interest is women who have gone off birth control in an attempt to become pregnant.

(c) We look in the total section of our two way table to find that out of the 678 women attempting to become pregnant, 244 succeeded in their first cycle, so $\hat{p} = 244/678 = 0.36$. For smokers we look only in the *Smoker* column of the two way table and observe 38 of 135 succeeded, so $\hat{p}_s = 38/135 = 0.28$. For non-smokers we look only in the *Non-smoker* column of the two way table and observe 206 of 543 succeeded, so $\hat{p}_{ns} = 206/543 = 0.38$.

(d) For the difference in proportions, we have $\hat{p}_{ns} - \hat{p}_s = 0.38 - 0.28 = 0.10$. This means that in this sample, the percent of non-smoking women successfully getting pregnant in the first cycle is 10 percentage points higher than the percent of smokers.

2.27 (a) The total number of respondents is 27,268 and the number answering zero is 18,712, so the proportion is $18712/27268 = 0.686$. We see that about 68.6% of respondents have not had five or more drinks in a single sitting at any time during the last two weeks.

(b) We see that 853 students answer five or more times and 495 of these are male, so the proportion is $495/853 = 0.580$. About 58% of those reporting that they drank five or more alcoholic drinks at least five times in the last two weeks are male.

(c) There are 8,956 males in the survey and $912 + 495 = 1407$ of them report that they have had five or more alcoholic drinks at least three times, so the proportion is $1407/8956 = 0.157$. About 15.7% of male college students report having five or more alcoholic drinks at least three times in the last two weeks.

(d) There are 18,312 females in the survey and $966 + 358 = 1324$ of them report that they have had five or more alcoholic drinks at least three times, so the proportion is $1324/18312 = 0.072$. About 7.2% of female college students report having five or more alcoholic drinks at least three times in the last two weeks.

2.29 (a) More females answered the survey since we see in graph (a) that the bar is much taller for females.

(b) It appears to be close to equal numbers saying they had no stress, since the height of the brown bars in graph (a) are similar. Graph (a) is the appropriate graph here since we are being asked about actual numbers not proportions.

(c) In this case, we are being asked about percents, so we use the relative frequencies in graph (b). We see in graph (b) that a greater percent of males said they had no stress.

(d) We are being asked about percents, so we use the relative frequencies in graph (b). We see in graph (b) that a greater percent of females said that stress had negatively affected their grades.

2.31 Graph (b) is the impostor. It shows more parochial students than private school students. The other three graphs have more private school students than parochial.

Section 2.2 Solutions

2.33 Only histogram F is skewed to the right.

2.35 While all of B,C,D,E and G are approximately symmetric, only B,C and E are also bell shaped.

2.37 Histograms E and G are both approximately symmetric, so the mean and median will be approximately equal. Histogram F is skewed right, so the mean should be larger then the median; while histogram H is skewed left, so the mean should be smaller then the median.

2.39 There are many possible dotplots we could draw that would be clearly skewed to the left. One is shown.

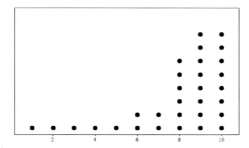

2.41 There are many possible dotplots we could draw that are approximately symmetric but not bell-shaped. One is shown.

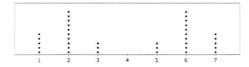

2.43 (a) We have $\bar{x} = (8 + 12 + 3 + 18 + 15)/5 = 11.2$.

   (b) The median is the middle number when the numbers are put in order smallest to largest. In order, we have:

$$3 \quad 8 \quad 12 \quad 15 \quad 18.$$

The median is $m = 12$. Notice that there are two data values less than the median and two data values greater.

   (c) There do not appear to be any particularly large or small data values relative to the rest, so there do not appear to be any outliers.

2.45 (a) We have $\bar{x} = (15 + 22 + 12 + 28 + 58 + 18 + 25 + 18)/8 = 24.5$.

(b) Since there are $n = 8$ values, the median is the average of the two middle numbers when the numbers are put in order smallest to largest. In order, we have:

$$12 \quad 15 \quad 18 \quad 18 \quad 22 \quad 25 \quad 28 \quad 58.$$

The median is the average of 18 and 22, so $m = 20$. Notice that there are four data values less than the median and four data values greater.

(c) The value 58 is significantly larger than all the other data values, so 58 is a likely outlier.

2.47 This is a sample, so the correct notation is $\bar{x} = 2386$ calories per day.

2.49 This is a population, so the correct notation is $\mu = 41.5$ yards per punt.

2.51 (a) We expect the mean to be larger since there appears to be a relatively large outlier (26.0) in the data values.

(b) There are eight numbers in the data set, so the mean is the sum of the values divided by 8. We have:

$$\text{Mean} = \frac{0.8 + 1.9 + 2.7 + 3.4 + 3.9 + 7.1 + 11.9 + 26.0}{8} = \frac{57.7}{8} = 7.2 \text{ mg/kg.}$$

The data values are already in order smallest to largest, and the median is the average of the two middle numbers. We have:

$$\text{Median} = \frac{3.4 + 3.9}{2} = 3.65.$$

2.53 (a) Since there are only 50 states and all of them are represented, this is the entire population.

(b) The distribution is skewed to the right. There appears to be an outlier at about 35 million. (The outlier represents the state of California.)

(c) The median splits the data in half and appears to be about 4 million. (In fact, it is exactly 4.170 million.)

(d) The mean is the balance point for the histogram and is harder to estimate. It appears to be about 6 million. (In fact, it is exactly 5.862 million.)

2.55 (a) The distribution is skewed to the left.

(b) The median is the value with half the area to the left and half to the right. The value 5 has way more area on the right so it cannot be correct. If we draw a line at 7, there is more area to the left than the right. The answer must be between 5 and 7 and a line at 6.5 appears to split the area into approximately equal amounts. The median is about 6.5.

(c) Because the data is skewed to the left, the values in the longer tail on the left will pull the mean down. The mean will be smaller than the median.

2.57 (a) We compute $\bar{x} = 26.6$. Since there are ten numbers, we average the two middle numbers to find the median. We have $m = (15 + 17)/2 = 16$.

(b) Without the outlier, we have $\bar{x} = 16.78$. Since $n = 9$, the median is the middle number. We have $m = 15$.

(c) The outlier has a very significant effect on the mean and very little effect on the median.

2.59 (a) Many people send just a few text messages per day, so many of the data values will be relatively small. However, some people send and receive a very large number of text messages per day. We expect that the bulk of the data values will be small numbers, with a tail extending out to the right to some values that are very large. This describes a distribution that is skewed to the right.

(b) The people with a very large number of text messages will pull up the mean but not the median, so we expect the mean to be 39.1 text messages and the median to be 10 messages. The fact that the mean and the median are so different indicates that the data is significantly skewed to the right, and almost certainly has some high outliers. In addition, the median of 10 messages implies that half the people averaged less than 10 text messages a day and half averaged more.

2.61 The notation for a median is m. We use m_H to represent the median earnings for high school graduates and m_C to represent the median earnings for college graduates. (You might choose to use different subscripts, which is fine.) The difference in medians is $m_H - m_C = 626 - 1025 = -399$. College graduates earn about $400 more per week than high school graduates.

2.63 (a) There are many possible answers. One way to force the outcome is to have a very small outlier, such as
2,   51,   52,   53,   54.
The median of these 5 numbers is 52 while the mean is 42.4.

(b) There are many possible answers. One way to force the outcome is to have a very large outlier, such as
2,   3,   4,   5,   200.
The median of these 5 numbers is 4 while the mean is 42.8.

(c) There are many possible answers. One option is the following:
2,   3,   4,   5,   6.
Both the mean and the median are 4.

2.65 The values are in order smallest to largest, and since more than half the values are 1, the median is 1. We calculate the mean to be $\bar{x} = 3.2$. In this case, the mean is probably a better value (despite the fact that 12 might be an outlier) since it allows us to see that some of the data values are above 1.

Section 2.3 Solutions

2.67   (a) Using technology, we see that the mean is $\overline{x} = 17.36$ with a standard deviation of $s = 5.73$.

   (b) Using technology, we see that the five number summary is $(10, 13, 17, 21, 28)$. Notice that these five numbers divide the data into fourths.

2.69   (a) Using technology, we see that the mean is $\overline{x} = 10.4$ with a standard deviation of $s = 5.32$.

   (b) Using technology, we see that the five number summary is $(4, 5, 11, 14, 22)$. Notice that these five numbers divide the data into fourths.

2.71   (a) Using technology, we see that the mean is $\overline{x} = 9.05$ hours per week with a standard deviation of $s = 5.74$.

   (b) Using technology, we see that the five number summary is $(0, 5, 8, 12, 40)$. Notice that these five numbers divide the data into fourths.

2.73 We know that the standard deviation is a measure of how spread out the data are, so larger standard deviations go with more spread out data. All of these histograms are centered at 10 and have the same horizontal scale, so we need only look at the spread. We see that $s = 1$ goes with Histogram B and $s = 3$ goes with Histogram C and $s = 5$ goes with Histogram A.

2.75 Remember that the five number summary divides the data (and hence the area in the histogram) into fourths.

   (a) II

   (b) V

   (c) IV

   (d) I

   (e) III

   (f) VI

2.77 The mean appears to be at about $\overline{x} \approx 500$. Since 95% of the data appear to be between about 460 and 540, we see that two standard deviations is about 40 so one standard deviation is about 20. We estimate the standard deviation to be between 20 and 25.

2.79 The minimum appears to be at 440, the median at 500, and the maximum at 560. The quartiles are a bit harder to estimate accurately. It appears that the lower quartile is about 485 and the upper quartile is about 515, so the five number summary is approximately $(440, 485, 500, 515, 560)$.

2.81 Since there are exactly $n = 100$ data points, the 10th-percentile is the value with 10 dots to the left of it. We see that this is at the value 62. Similarly, the 90th-percentile is the value with 10 dots to the right of it. This is the value 73.

2.83 This dataset is very symmetric.

2.85 For this dataset, half of the values are clustered between 22 and 27, and the other half are spread out to the left all the way down to 0. This distribution is skewed to the left.

2.87 We have
$$Z\text{-score} = \frac{\text{Value} - \text{Mean}}{\text{Standard deviation}} = \frac{243 - 200}{25} = 1.72.$$
This value is 1.72 standard deviations above the mean.

2.89 We have
$$Z\text{-score} = \frac{\text{Value} - \text{Mean}}{\text{Standard deviation}} = \frac{5.2 - 12}{2.3} = -2.96.$$
This value is 2.96 standard deviations below the mean, which is quite extreme in the lower tail.

2.91 The 95% rule says that 95% of the data should be within two standard deviations of the mean, so the interval is:

$$
\begin{array}{ccl}
\text{Mean} & \pm & 2 \cdot \text{StDev} \\
200 & \pm & 2 \cdot (25) \\
200 & \pm & 50 \\
150 & \text{to} & 250.
\end{array}
$$

We expect 95% of the data to be between 150 and 250.

2.93 The 95% rule says that 95% of the data should be within two standard deviations of the mean, so the interval is:

$$
\begin{array}{ccl}
\text{Mean} & \pm & 2 \cdot \text{StDev} \\
1000 & \pm & 2 \cdot (10) \\
1000 & \pm & 20 \\
980 & \text{to} & 1020.
\end{array}
$$

We expect 95% of the data to be between 980 and 1020.

2.95 (a) The numbers range from 46 to 61 and seem to be grouped around 53. We estimate that $\overline{x} \approx 53$.

(b) The standard deviation is roughly the typical distance of a data value from the mean. All of the data values are within 8 units of the estimated mean of 53, so the standard deviation is definitely not 52, 10, or 55. A typical distance from the mean is clearly greater than 1, so we estimate that $s \approx 5$.

(c) Using a calculator or computer, we see $\overline{x} = 52.875$ and $s = 5.07$.

2.97 (a) We see in the computer output that the five number summary is (17.800, 22.175, 24.400, 26.825, 30.900).

(b) The range is the difference between the maximum and the minimum, so we have Range $= 30.9 - 17.8 = 13.1$. The interquartile range is the difference between the first and third quartiles, so we have $IQR = 26.825 - 22.175 = 4.65$.

(c) The 15th-percentile is between the minimum and the first quartile, so it will be between 17.8 and 22.175. The 60th-percentile is between the median and the third quartile, so it will be between 24.4 and 26.825.

2.99 (a) See the table.

Year	Joey	Takeru	Difference
2009	68	64	4
2008	59	59	0
2007	66	63	3
2006	52	54	-2
2005	32	49	-17

(b) For the five differences, we use technology to see that the mean is -2.4 and the standard deviation is 8.5.

2.101   (a) We expect that 95% of the data will lie between $\overline{x} \pm 2s$. In this case, the mean is $\overline{x} = 2.31$ and the standard deviation is $s = 0.96$, so 95% of the data lies between $2.31 \pm 2(0.96)$. Since $2.31 - 2(0.96) = 0.39$ and $2.31 + 2(0.96) = 4.23$, we estimate that about 95% of the temperature increases will lie between $0.39°$ and $4.23°$.

(b) Since $\overline{x} = 2.31$ and $s = 0.96$, the z-score for a temperature increase of $4.9°$ is

$$z\text{-score} = \frac{x - \overline{x}}{s} = \frac{4.9 - 2.31}{0.96} = 2.70.$$

The temperature increase for this man is 2.7 standard deviations above the mean.

2.103   (a) Using software, we see that $\overline{x} = 0.272$ and $s = 0.237$.

(b) The largest concentration is 0.851. The z-score is

$$z\text{-score} = \frac{x - \overline{x}}{s} = \frac{0.851 - 0.272}{0.237} = 2.44.$$

The largest value is almost two and a half standard deviations above the mean, and appears to be an outlier.

(c) Using software, we see that

$$\text{Five number summary} = (0.073, 0.118, 0.158, 0.358, 0.851).$$

(d) The range is $0.851 - 0.073 = 0.778$ and the interquartile range is $IQR = 0.358 - 0.118 = 0.240$.

2.105 The data is not at all bell-shaped, so it is not appropriate to use the 95% rule with this data.

2.107 We first calculate the z-scores for the four values. In each case, we use the fact that the z-score is the value minus the mean, divided by the standard deviation. We have

$$z\text{-score for } FGPct = \frac{0.510 - 0.464}{0.053} = 0.868 \qquad z\text{-score for } Points = \frac{2111 - 994}{414} = 2.698$$
$$z\text{-score for } Assists = \frac{554 - 220}{170} = 1.965 \qquad z\text{-score for } Steals = \frac{124 - 68.2}{31.5} = 1.771$$

The most impressive statistic is his total points, which is about 2.7 standard deviations above the mean. The least impressive is his field goal percentage, which is only 0.868 standard deviations above the mean. He is above the mean on all four, however, and substantially above the mean on three.

2.109   (a) Using technology, we find that the mean is $\overline{x} = 65.89$ percent with a fast connection and the standard deviation is $s = 18.29$.

(b) Using technology, we find that the mean is $\overline{x} = 26.18$ hours online and the standard deviation is $s = 3.41$.

(c) It is not clear if there is a relationship or not. The Swiss have the fastest connection times and the lowest time online and Brazil has the slowest connection time and the highest time online, but the pattern does not seem to be obvious with the other countries.

2.111 (a) Using technology, we see that the mean is $\overline{x} = 13.15$ years with a standard deviation of $s = 7.24$ years.

(b) We have

$$\text{The } z\text{-score for the elephant } = \frac{\text{Elephant's value} - \text{Mean}}{\text{Standard deviation}} = \frac{40 - 13.15}{7.24} = 3.71.$$

The elephant is 3.71 standard deviations above the mean, which is way out in the upper tail of the distribution. The elephant is a strong outlier!

2.113 (a) The smallest possible standard deviation is zero, which is the case if the numbers don't deviate at all from the mean. The dataset is:
5, 5, 5, 5, 5, 5.

(b) The largest possible standard deviation occurs if all the numbers are as far as possible from the mean of 5. Since we are limited to numbers only between 1 and 9 (and since we have to keep the mean at 5), the best we can do is three values at 1 and three values at 9. The dataset is:
1, 1, 1, 9, 9, 9.

2.115 A bell-shaped distribution with mean 7 and standard deviation 1.

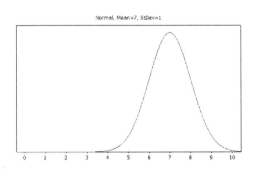

2.117 A bell-shaped distribution with mean 5 and standard deviation 0.5.

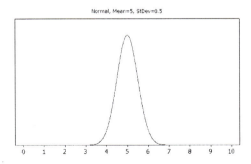

2.119   (a) The rough estimate is $(130 - 35)/5 = 19$ bpm compared to the actual standard deviation of s=12.2 bpm. The possible outliers pull the rough estimate up quite a bit. Without the two outliers, the rough estimate is $(96 - 35)/5 = 12.2$, exactly matching the actual standard deviation.

   (b) The rough estimate is $(40 - 0)/5 = 8$. The rough estimate is a bit high compared to the actual standard deviation of $s = 5.741$.

   (c) The rough estimate is $(40 - 1)/5 = 7.8$. The rough estimate is quite close to the actual standard deviation of $s = 7.24$.

Section 2.4 Solutions

2.121 We match the five number summary with the maximum, first quartile, median, third quartile, and maximum shown in the boxplot.

   (a) This five number summary matches boxplot W.

   (b) This five number summary matches boxplot X.

   (c) This five number summary matches boxplot Y.

   (d) This five number summary matches boxplot Z.

2.123   (a) Half of the data appears to lie in the small area between 20 and 40, while the other half (the right tail) appears to extend all the way up from 40 to about 140. This distribution appears to be skewed to the right.

   (b) Since there are no asterisks on the graph, there are no outliers.

   (c) The median is approximately 40 and since the distribution is skewed to the right, we expect the values out in the right tail to pull the mean up above the median. A reasonable estimate for the mean is about 50.

2.125   (a) This distribution isn't perfectly symmetric but it is close. Despite the presence of outliers on both sides, there does not seem to be a distinct skew either way. This distribution is approximately symmetric.

   (b) There appear to be 3 low outliers and 2 high outliers.

   (c) Since the distribution is approximately symmetric, we expect the mean to be close to the median, which appears to be at about 1200. We estimate that the mean is about 1200.

2.127   (a) We see that $Q_1 = 42$ and $Q_3 = 56$ so the interquartile range is IQR $= 56 - 42 = 14$. We compute

$$Q_1 - 1.5(IQR) = 42 - 1.5(14) = 21,$$

and

$$Q_3 + 1.5(IQR) = 56 + 1.5(14) = 77.$$

There are two data values that fall outside these values. We see that 15 and 20 are both small outliers.

   (b) Notice that the line on the left of the boxplot extends down to 28, the smallest data value that is not an outlier.

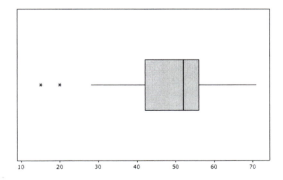

2.129   (a) We see that $Q_1 = 10$ and $Q_3 = 16$ so the interquartile range is IQR $= 16 - 10 = 6$. We compute

$$Q_1 - 1.5(IQR) = 10 - 1.5(6) = 1,$$

and

$$Q_3 + 1.5(IQR) = 16 + 1.5(6) = 25.$$

There are no small outliers and there are two large outliers, at 28 and 30.

(b) Notice that the line on the right extends up to 23, the largest data value that is not an outlier.

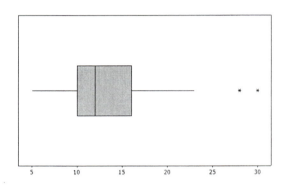

2.131 We see from the five number summary that the interquartile range is $IQR = Q_3 - Q_1 = 77 - 49 = 28$. Using the IQR, we compute:

$$\begin{aligned} Q_1 - 1.5 \cdot IQR &= 49 - 1.5(28) = 49 - 42 = 7. \\ Q_3 + 1.5 \cdot IQR &= 77 + 1.5(28) = 77 + 42 = 119. \end{aligned}$$

Scores greater than 119 are impossible since the scale only goes up to 100. An audience score less than 7 would qualify as a low outlier. (That would have to be a *very* bad movie!) Since the minimum rating (seen in the five number summary) is 24, there are no low outliers.

2.133   (a) The highest mean is in Drama, while the lowest mean is in Horror movies.

(b) The highest median is in Drama, while the lowest median is in Action movies.

(c) The lowest score is 24 and it is for a Thriller. The highest score is 93, and it is obtained by both an Action and a Comedy.

(d) The genre with the largest number of movies is Action, with $n = 32$.

2.135   (a) We see that the interquartile range is $IQR = Q_3 - Q_1 = 149 - 15 = 134$. We compute:

$$Q_1 - 1.5(IQR) = 15 - 1.5(134) = 15 - 201 = -186.$$

and

$$Q_3 + 1.5(IQR) = 149 + 1.5(134) = 149 + 201 = 350.$$

Outliers are any values outside these fences. In this case, there are four outliers that are larger than 350. The four outliers are 402, 447, 511, and 536.

(b) A boxplot of time to infection is shown:

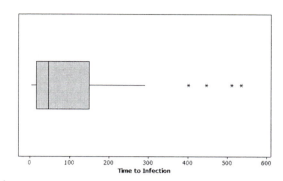

2.137   (a) We have IQR $= 2106 - 1334 = 772$, so the upper boundary for non-outlier data values is:

$$\begin{aligned} Q_3 + 1.5(IQR) & = 2106 + 1.5(772) \\ & = 2106 + 1158 \\ & = 3264. \end{aligned}$$

Any data value above 3264 is an outlier, so the seven largest calorie counts are all outliers.

(b) We have already seen that $IQR = 772$, so the lower boundary for non-outlier data values is

$$\begin{aligned} Q_1 - 1.5(IQR) & = 1334 - 1.5(772) \\ & = 1334 - 1158 \\ & = 176. \end{aligned}$$

We see in the five number summary that the minimum data value is 445, so there are no values below 176 and no low outliers.

(c) A boxplot of daily calorie consumption is shown:

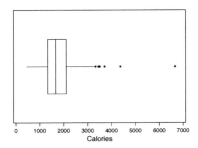

2.139 The side-by-side boxplots are almost identical. Vitamin use appears to have no effect on the concentration of retinol in the blood.

2.141 Both distributions are relatively symmetric with one or two outliers. In general, the blood pressures of patients who lived appear to be slightly higher as a group than those of the patients who died. The middle 50% box for the surviving patients is shifted to the right of the box for patients who died and shows a smaller interquartile range. Both quartiles and the median are larger for the surviving group. Note that the boxplots give no information about how many patients are in each group. From the original data table, we can find that 40 of the 200 patients died and the rest survived.

2.143   (a) See the figure. It appears that respiration rate is higher when calcium levels are low, and that there is not much difference in respiration rate between medium and high levels of calcium.

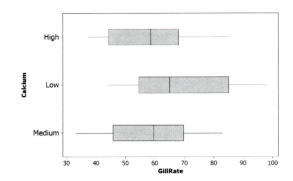

(b) With a low calcium level, the mean is 68.50 beats per minute with a standard deviation of 16.23. With a medium level, the mean is 58.67 beats per minute with a standard deviation of 14.28. With a high level of calcium, the mean is 58.17 beats per minute with a standard deviation of 13.78. Again, we see that respiration is highest with low calcium.

(c) This is an experiment since the calcium level was actively manipulated.

2.145 Here is one possible graph of the side-by-side boxplots:

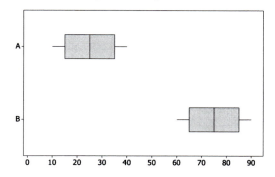

2.147 Answers will vary.

Section 2.5 Solutions

2.149 A correlation of -1 means the points all lie exactly on a line and there is a negative association. The matching scatterplot is (b).

2.151 A correlation of 0.8 means there is an obvious positive linear association in the scatterplot, but there is some deviation from a perfect straight line. The matching scatterplot is (d).

2.153 The correlation represents almost no linear relationship, so the matching scatterplot is (c).

2.155 The correlation shows a strong positive linear association, so the matching scatterplot is (d).

2.157 We expect that larger houses will cost more to heat, so we expect a positive association.

2.159 We wear more clothes when it is cold outside, so as the temperature goes down, the amount of clothes worn goes up. This describes a negative association.

2.161 Usually there are not many people in a heavily wooded area and there are not many trees in a heavily populated area such as the middle of a city. This would be a negative relationship.

2.163 See the figure below.

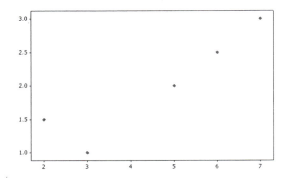

2.165 The correlation is $r = 0.915$.

2.167 (a) The incumbent lost 3 times and won 8 times. We have $3/11 = 0.273$. Since 1940, the sitting president has lost his bid for re-election 27.3% of the time.

(b) President Eisenhower had the highest approval rating at 70%. President Nixon had the highest margin of victory at 23.2%.

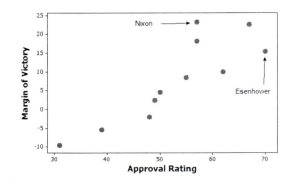

2.169 The explanatory variable is roasting time and the response variable is the amount of caffeine. The two variables have a negative association.

2.171 Since more cheating by the mother is associated with more cheating by the daughter, this is a positive association.

2.173 (a) There appears to be a negative association, which means in this context that states with a larger proportion of the population eating lots of vegetables tend to have a lower obesity rate. This makes sense since people who eat lots of vegetables are less likely to be obese.

  (b) A healthy state would include a large percentage of people eating lots of vegetables and a small percentage of people who are obese, which corresponds to the bottom right corner. An unhealthy state would include a small percentage of people eating lots of vegetables and a high obesity rate, which corresponds to the top left corner.

  (c) There is one point furthest in the bottom right corner and it corresponds to the state of Vermont. There are three dots in the unhealthy top left corner, and they correspond to the states of Mississippi, Kentucky, and Oklahoma.

  (d) Since all 50 US States are included, this is a population. The correct notation is ρ.

  (e) The two variables appear to be negatively correlated with a moderately strong linear relationship. The correct correlation is $\rho = -0.605$.

  (f) No, correlation does not imply cause and effect relationships.

  (g) No, correlation does not imply cause and effect relationships.

  (h) That point corresponds to the state of Colorado.

2.175 (a) There are three variables mentioned: how closed a person's body language is, level of stress hormone in the body, and how powerful the person felt. Since results are recorded on numerical scales that represent a range for body language and powerful, all three variables are quantitative.

  (b) People with a more closed posture (low values on the scale) tended to have higher levels of stress hormones, so there appears to be a negative relationship. If the scale for posture had been reversed, the answer would be the opposite. A positive or negative relationship can depend on how the data is recorded.

  (c) People with a more closed posture (low values on the scale) tended to feel less powerful (low values on that scale), so there appears to be a positive relationship. If both scales were reversed, the answer would not change. If only one of the scales was reversed, the answer would change.

2.177 (a) A positive relationship would imply that a student who exercises lots also watches lots of television, and a student who doesn't exercise also doesn't watch much TV. A negative relationship implies that students who exercise lots tend to not watch much TV and students who watch lots of TV tend to not exercise much.

  (b) A student in the top left exercises lots and watches very little television. A student in the top right spends lots of hours exercising and also spends lots of hours watching television. (Notice that there are no students in this portion of the scatterplot.) A student in the bottom left does not spend much time either exercising or watching television. (Notice that there are lots of students in this corner.) A student in the bottom right watches lots of television and doesn't exercise very much.

  (c) The outlier on the right watches a great deal of television and spends very little time exercising. The outlier on the top spends a great deal of time exercising and watches almost no television.

(d) There is essentially no linear relationship between the number of hours spent exercising and the number of hours spent watching television.

2.179   (a) Here is a scatterplot of Jogger A vs Jogger B.

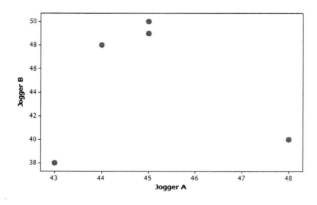

(b) The correlation between the two joggers is -0.096.

(c) The correlation between the two joggers with the windy race added is now 0.562.

(d) Adding the results from the windy day has a very strong effect on the relationship between the two joggers!

2.181 Type of liquor is a categorical variable, so correlation should not be computed and a positive relationship has no meaning. There cannot be a *linear* relationship involving a categorical variable.

2.183 There are many ways to draw the scatterplots. One possibility for each is shown in the figure.

2.185   (a) Since we are looking to see if budget affects audience score, we put the explanatory variable (budget) on the horizontal axis and the response variable (audience score) on the vertical axis. See the figure.

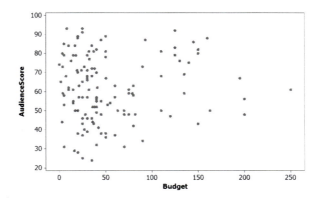

(b) There is not a strong linear relationship. In context, this means that having a larger budget for a movie does not tend to make for a higher (or lower) score from the audience.

(c) The outlier has a budget about 250 million dollars. This movie is *Pirates of the Caribbean* and the audience score is 61. The movie with a budget of 125 million dollars and an audience score over 90 is *Harry Potter and the Deathly Hallows, Part 2*.

(d) We use technology to see that the correlation is 0.084. This correlation near 0 matches the lack of an association we see in the scatterplot.

Section 2.6 Solutions

2.187   (a) The predicted value for the data point is $\widehat{Hgt} = 24.3 + 2.74(12) = 57.18$ inches. The residual is $60 - 57.18 = 2.82$. This child is 2.82 inches taller than the predicted height.

(b) The slope 2.74 tells the expected change in Hgt given a one year increase in Age. We expect a child to grow about 2.74 inches per year.

(c) The intercept 24.3 tells the Hgt when the Age is 0, or the height (or length) of a newborn. This context does make sense, although the estimate is rather high.

2.189   (a) The predicted value for the data point is $\widehat{Weight} = 95 + 11.7(5) = 153.5$ lbs. The residual is $150 - 153.5 = -3.5$. This individual is capable of bench pressing 3.5 pounds less than predicted.

(b) The slope 11.7 tells the expected change in Weight given a one hour a week increase in Training. If an individual trains an hour more each week, the predicted weight the individual is capable of bench pressing would go up 11.7 pounds.

(c) The intercept 95 tells the Weight when the hours Training is 0, or the bench press capability of an individual who never lifts weights. This intercept does make sense in context.

2.191   The regression equation is $\hat{Y} = 0.395 + 0.349X$.

2.193   The regression equation is $\hat{Y} = 111.7 - 0.84X$.

2.195   (a) Year is the explanatory variable and CO_2 concentration is the response variable.

(b) A scatterplot of CO_2 vs $Year$ is shown. There is a very strong linear relationship in the data.

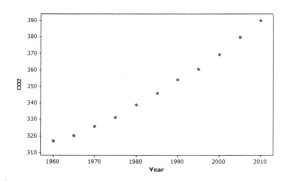

(c) We find that $r = 0.993$. This correlation is very close to 1 and matches the very strong linear relationship we see in the scatterplot.

(d) We see that $\widehat{CO_2} = -2571 + 1.47(Year)$.

(e) The slope is 1.47. Carbon dioxide concentrations in the atmosphere have been going up at a rate of about 1.47 ppm each year.

(f) The intercept is -2571. This is the expected CO_2 concentration in the year 0, but clearly doesn't make any sense since the concentration can't be negative. The linear trend clearly does not extend back that far and we can't extrapolate back all the way to the year 0.

(g) In 2003, the predicted CO_2 concentration is $\widehat{CO_2} = -2571 + 1.47(2003) = 373.41$. This seems reasonable since the value lies between the data values for years 2000 and 2005.
In 2020, the predicted CO_2 concentration is $\widehat{CO_2} = -2571 + 1.47(2020) = 398.4$. We have less confidence in this prediction since we can't be sure the linear trend will continue.

(h) In 2010, the predicted CO_2 concentration is $\widehat{CO_2} = -2571 + 1.47(2010) = 383.7$. We see in the data that the actual concentration that year is 389.78, so the residual is $389.78 - 383.7 = 6.08$. The CO_2 concentration in 2010 was quite a bit above the predicted value.

2.197 (a) The trend is positive, with a mild linear relationship.

(b) The residual is the vertical distance from the point to the line, which is much larger in 2007 than it is in 2008. We see that in 2010, the point is below the line so the observed value is less than the predicted value. The residual is negative.

(c) We use technology to see that the correlation is $r = 0.692$.

(d) We use technology to find the regression line: $\widehat{HotDogs} = -3385 + 1.72 \cdot Year$.

(e) The slope indicates that the winning number is going up by about 1.72 more hot dogs each year. People better keep practicing!

(f) The predicted number in 2012 is $\widehat{HotDogs} = -3385 + 1.72 \cdot (2012) = 75.64$, which is a very large number of hot dogs!

(g) It is not appropriate to use this regression line to predict the winning number in 2020 because that is extrapolating too far away from the years that were used to create the dataset.

2.199 The slope is 0.836. The slope tells us the expected change in the response variable (Margin) given a one unit increase in the predictor variable (Approval). In this case, we expect the margin of victory to go up by 0.836 if the approval rating goes up by 1.
The y-intercept is -36.5. This intercept tell us the expected value of the response variable (Margin) when the predictor variable (Approval) is zero. In this case, we expect the margin of victory to be -36.5 if the approval rating is 0. In other words, if *no one* approves of the job the president is doing, the president will lose in a landslide. This is not surprising!

2.201 The man with the largest positive residual weighs about 190 pounds and has a body fat percentage about 40%. The predicted body fat percent for this man is about 20% so the residual is about $40 - 20 = 20$.

2.203 (a) For 35 cm, the predicted body fat percent is $\widehat{BodyFat} = -47.9 + 1.75 \cdot (35) = 13.35\%$. For a neck circumference of 40 cm, the predicted body fat percent is $\widehat{BodyFat} = -47.9 + 1.75 \cdot (40) = 22.1\%$.

(b) The slope of 1.75 indicates that as neck circumference goes up by 1 cm, body fat percent goes up by 1.75.

(c) The predicted body fat percent for this man is $\widehat{BodyFat} = -47.9 + 1.75 \cdot (38.7) = 19.825\%$, so the residual is $11.3 - 19.825 = -8.525$.

2.205 (a) We are attempting to predict rural population with land area, so land area is the explanatory variable, and percent rural is the response.

(b) There appears to be some positive correlation between these two variables, so the most likely correlation is 0.60.

(c) Using technology the regression line is: $\widehat{Rural} = 28.99 + 0.079(LandArea)$. The slope is 0.079, which means percent rural goes up by about 0.079 with each increase in 1,000 sq km of country size.

(d) The intercept does not make sense, since a country of size zero would have no population at all!

(e) The most influential country is the one in the far top right, which is Uzbekistan (UZB). This is due to the fact that Uzbekistan is much larger then any of the other countries sampled, so it appears to be an outlier for the explanatory variable.

(f) Predicting the percent rural for USA with the prediction equation gives $\widehat{Rural} = 28.99 + 0.079(9147.4) = 751.63$. This implies that 752% of the United States population lives in rural areas, which doesn't make any sense at all. The regression line does not work at all for the US, because we are extrapolating so far outside of the original sample of 10 land area values. The US is much larger in area than any of the countries that happened to be picked for the random sample.

2.207 (a) See the figure. We see that there is a relatively strong positive linear trend. It appears that the opening weekend is a reasonably good predictor of future total world earnings for a movie.

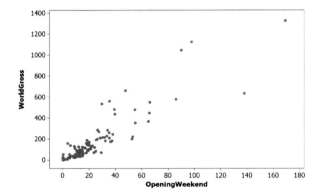

(b) The movie is *Harry Potter and the Deathly Hallows, Part 2*.

(c) We find that the correlation is $r = 0.904$.

(d) The regression line is $\widehat{WorldGross} = -8.7 + 7.70 \cdot OpeningWeekend$.

(e) If a movie makes 50 million dollars in its opening weekend, we predict that total world earnings for the year for the movie will be $\widehat{WorldGross} = -8.7 + 7.70 \cdot (50) = 376.3$ million dollars.

2.209 Answers will vary.

Unit A: Essential Synthesis Solutions

A.1 (a) This is an experiment since subjects were randomly assigned to one of three groups which determined what method was used.

(b) This study could not be "blind" since both the participants and those recording the results could see what each had applied.

(c) The sample is the 46 subjects participating in the experiment. The intended population is probably anyone who might consider using black grease under the eyes to cut down on glare from the sun.

(d) One variable is the improvement in contrast sensitivity, and this is a quantitative variable. A second variable records what group the individual is in, and this is a categorical variable.

(e) Since we are examining the relationship between a categorical variable and a quantitative variable, we could use side-by-side boxplots to display the results.

A.3 (a) The cases are the students (or the students' computers). The sample size is $n = 45$. The sample is not random since the students were specifically recruited for the study.

(b) This is an observational study, since none of the variables was actively manipulated.

(c) For each student, the variables recorded are: number of active windows per lecture, percent of windows that are distracting, percent of time on distracting windows, and score on the test of the material. All of these variables are quantitative. (Note whether or not a window is distracting is categorical, but for each student the *percentage* of distracting windows is quantitative.)

(d) The number of active windows opened per lecture is a single quantitative variable, so we might use a histogram, dotplot, or boxplot. If we want outliers clearly displayed, a boxplot would be the best choice.

(e) The association described is between two quantitative variables, so we would use a scatterplot. An appropriate statistic would be a correlation. Since more time on distracting websites is associated with lower test scores, it is a negative association.

(f) No, we cannot conclude that the time spent at distracting sites causes lower test scores, since this is an observational study not an experiment. There are many possible confounding variables and we cannot infer a cause and effect relationship (although there might be one).

(g) We consider the time on distracting websites the explanatory variable, and the exam score the response variable.

(h) To make this cause and effect conclusion, we would need to do a randomized experiment. One option would be to randomly divide a group of students into two groups and have one group use distracting websites and the other not have access to such websites. Compare test scores of the two groups at the end of the study. It is probably not feasible to require one group to visit distracting websites during class!

A.5 One design would be to divide our human subjects into two groups. We would control the amount of calories so that it is similar between the two groups. We would also control when the calories were consumed: One group to eat most of their calories during the day and one group to eat most of their calories at night. After a reasonable amount of time, we measure weight gain in all subjects and compare the two groups. Other experimental designs are also possible.

A.7 (a) The sample is the 86 patients in the study. The intended population is all people with bladder cancer.

(b) One variable records whether or not the tumor recurs, and one records which treatment group the patient is in. Both variables are categorical.

(c) This is an experiment since treatments were randomly assigned. Since the experiment was double-blind, we know that neither the participants nor the doctors checking for new tumors knew who was getting which treatment.

(d) Since we are looking at a relationship between two categorical variables, it is reasonable to use a two-way table to display the data. The categories for treatment are "Placebo" and "Thiotepa" and the categories for outcome are "Recurrence" and "No Recurrence". The two-way table is shown.

	Recurrence	No recurrence	Total
Placebo	29	19	48
Thiotepa	18	20	38
Total	47	39	86

(e) We compare the proportion of patients for whom tumors returned between the two groups. For the placebo group, the tumor returned in $29/48 = 0.604 = 60.4\%$ of the patients. For the group taking the active drug, the tumor returned in $18/38 = 0.474 = 47.4\%$ of the patients. The drug appears to be more effective than the placebo, since the rate of recurrence of the tumors is lower for the patients in the thiopeta treatment group than for patients in the placebo group.

A.9 Answers will vary.

Unit A: Review Exercise Solutions

A.11 (a) The sample is the 200 patients from whom data were collected. A reasonable population is all patients admitted to the ICU at that hospital. Other answers are possible.

(b) The quantitative variables are: *Age*, *Systolic* (Systolic blood pressure), and *HeartRate*. The other 17 variables are all categorical.

(c) There are many possible answers, such as "What is the average age of a person being admitted to the ICU?" or "What proportion of patients admitted to the ICU survive?"

(d) There are many possible answers, such as "Does gender impact the likelihood of CPR being administered?" or "Is there a strong relationship between heart rate and age?"

A.13 (a) It is an observational study. The researcher asked the boys how often they ate fish and collected data on their intelligence test scores, but did nothing to change or determine their levels of fish consumption or intelligence.

(b) The explanatory variable is whether or not fish is consumed at least once a week, and the response variable is the score on the intelligence test.

(c) One possible confounding variable is the intelligence level of the parents. Families in which the parents are more intelligent may tend to eat more fish and also to have sons who score higher on an intelligence test. Other possible confounding variables might be whether boys live near the coast or inland, or how often the boys' parents provide homecooked meals. You can probably think of other possibilities. Remember that a confounding variable is a variable that might influence *both* the explanatory and the response variables.

(d) No. Observational studies cannot yield causal conclusions.

A.15 The article is assuming causation when it should not be, since the results come from an observational study. A possible confounding variable is the health of the men when they were tested at age 70. Poor health would cause slower walking *and* greater risk of death.

A.17 (a) The cases are the students and the sample size is 70.

(b) There are four variables mentioned. One is the treatment group (walk in sync, walk out of sync, or walk any way). The other three are the quantitative ratings given on the three questions of closeness, liking, and similarity.

(c) This is an experiment since the students were actively told how to follow the accomplice.

(d) There are many possible ways to draw this; one is shown in the figure.

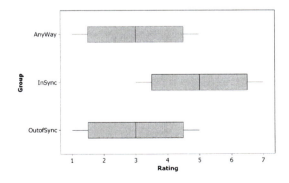

(e) One possible graph to use to look at a relationship of number of pill bugs killed by which treatment group the student was in would be a side-by-side boxplot, since one of these variables is quantitative and one is categorical. We could also use comparative dotplots or comparative histograms.
We would use a scatterplot to look at the association of number of pill bugs killed with the rating given on the liking accomplice scale, since both of these are quantitative variables.

A.19 Answers will vary. One possible sample is shown below.

$$7692, \ 1708, \ 0099, \ 4755, \ 1406, \ 4937, \ 6647, \ 2496, \ 3850, \ 4673$$

See the technology notes to see how to use specific technology to select a random sample.

A.21   (a) This is an experiment since facial features were actively manipulated.

(b) This "blinding" allows us to get more objective reactions to the video clips.

(c) We use $\overline{x}_S$ for the mean of the smiling group and $\overline{x}_N$ for the mean of the non-smiling group. The difference in means is $\overline{x}_S - \overline{x}_N = 7.8 - 5.9 = 1.9$.

(d) Since the results come from a randomized experiment, a substantial difference in the mean ratings would imply that smiling causes an increase in positive emotions.

A.23   (a) The sample is 48 participants. The population of interest is all people. The variable is whether or not each person's lie is detected.

(b) The proportion of time the lie detector fails to report deception is $\hat{p} = 17/48 = 0.35$.

(c) Since the lie detector fails 35% of the time, it is probably not reasonable to use it.

A.25 The histogram is relatively symmetric and bell-shaped. The mean appears to be approximately $\overline{x} = 7$. To estimate the standard deviation, we estimate an interval centered at 7 that contains approximately 95% of the data. The interval from 3 to 11 appears to contain almost all the data. Since 3 and 11 are both 4 units from the mean of 7, we have $2s = 4$, so the standard deviation appears to be approximately $s = 2$.

A.27   (a) There appear to be some low outliers pulling the mean well below the median. Half of the growing seasons over the last 50 years have been longer than 275 days and half have been shorter. Some of the growing seasons have been extremely short and have pulled the mean down to 240 days.

(b) Here is a smooth curve that could represent this distribution.

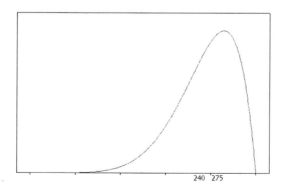

(c) The distribution is skewed to the left.

A.29 (a) The sample is the 1,917 people who filled out the survey. The intended population appears to be all cell phone users.

(b) Since percentages are given rather than the actual counts, this is a relative frequency table.

(c) The distribution is skewed to the right. The percentages are high for the categories with fewer calls (1-5 and 6-10) and get smaller as the number of calls get larger, even though the widths of the intervals increase. We would expect almost no tail on the left and a long tail to the right.

(d) The median is $m = 5.00$ and the mean is $\bar{x} = 13.10$. The strong right skew and a few large values greater than 30 pull up the mean and have little effect on the median.

A.31 (a) For most people, the vast majority of phone calls made are quite short. On the other hand, there are often a few very long phone calls during a month. We expect that the bulk of the data values will be between 0 and 5 minutes, with a tail extending out to the right to some phone calls extending perhaps as long as two hours (120 minutes). This describes a distribution that is skewed to the right.

(b) The extremely long phone calls will pull up the mean but not the median, so we expect the mean to be 13.7 minutes and the median to be 2.5 minutes. Notice that this implies that half the phone calls made on this cell phone are less than 2.5 minutes in length.

A.33 There were $67000/2 = 33500$ patients receiving each drug. Since $(0.587)(33500) = 19664.5$, we estimate that the number of people receiving paricalcitol who survived is about 19,665. Since $(0.515)(33500) = 17252.5$, we estimate that the number of people receiving calcitriol who survived is about 17,253. The total number of survivors is $19665 + 17253 = 36918$, The percent of survivors receiving paricalcitol is

$$\hat{p} = \frac{19665}{36918} = 53.3\%.$$

A two-way table (subject to round off error) is given below.

	Survived	Died	Total
Paricalcitol	19665	13835	33500
Calcitriol	17253	16247	33500
Total	36918	30082	67000

A.35 (a) There are many possible answers. One boxplot with a right skew is given.

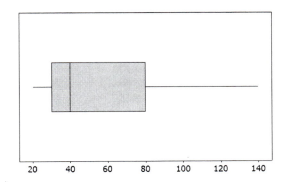

(b) There are many possible answers. One boxplot with a left skew is given.

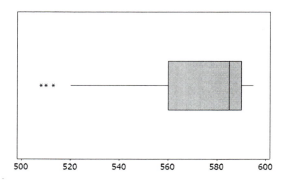

(c) There are many possible answers. One boxplot with a symmetric distribution is given.

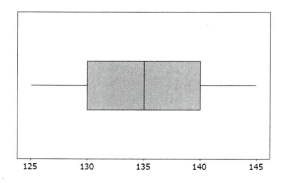

A.37 The five number summary for teens is $(100, 104, 130, 140, 156)$, so the IQR is 36. For teens typical values should fall between $104 - 1.5(36) = 50$ and $140 + 1.5(36) = 194$. Since all 13 blood pressure values for the teens are between these values, we find no outliers for that group.

The five number summary for patients in their eighties is $(80, 110, 135, 141, 190)$, so the IQR is 31. Typical values for the patients in their eighties fall between $110 - 1.5(31) = 63.5$ and $141 + 1.5(31) = 187.5$. This identifies the two blood pressures at 190 as unusually high values for eighty year old patients in this intensive care unit. These two values are both outliers.

A.39 (a) The mean is the balance point of the histogram and appears to be at approximately 100 beats per minute. Since the distribution is relatively symmetric and bell-shaped, we expect 95% of the data to be within two standard deviations of the mean. It appears that about 95% of the data is between about 50 and 150, so we estimate that the standard deviation is about 25. (In fact, the exact mean and standard deviation are $\bar{x} = 98.9$ and $s = 26.8$.)

(b) The 10th-percentile is the data point with 10% of the area of histogram boxes below it. It appears on the histogram that the 10th-percentile is about 60.

(c) The smallest heart rate appears to be about 40 and the largest appears to be about 190, so the range is roughly $190 - 40 = 150$.

A.41 (a) A scatterplot of verbal vs math SAT scores is shown below with the regression line (which is almost perfectly flat).

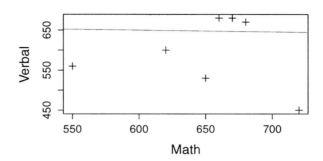

(b) Using technology we find the correlation between math and verbal scores for this sample is $r = -0.071$.

(c) Based on this small sample of seven pairs, computing a regression line to help predict verbal scores based on math scores is not very useful. The flat line in the scatterplot shows no consistent positive or negative linear trend between these variables. This is also seen with the sample correlation ($r = -0.071$) which is very close to zero.

A.43 (a) The scatterplot for the original five data points is shown below.

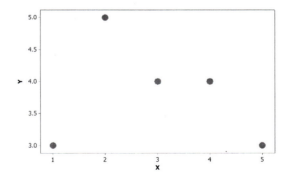

(b) We use technology to find $r = -0.189$. Both the scatterplot and the correlation show almost no linear relationship between x and y.

(c) The scatterplot with the extra point added is shown below.

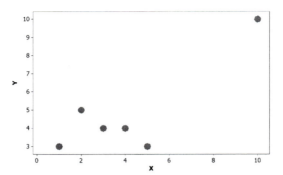

(d) The correlation (with the extra data point) is $r = 0.836$.

(e) When the outlier $(10, 10)$ is added, the correlation is suddenly very strong, and even changes from negative to positive. The outlier has a very substantial effect on the correlation.

A.45 (a) Both correlations appear to be positive.

(b) The linear relationship is stronger for the calories and fat data, so the correlation will be higher for these two variables.

(c) We locate this individual point as the only point above the 4000 calorie mark on the vertical scale. The fat consumption appears to be about 240 grams, which is an extreme high value for fat. The fiber consumption for this individual appears to be about 23 grams of fiber, which is not an extreme value for fiber consumption.

A.47 (a) The slope 0.654 means that for a one percent rise in the high school graduation rate, the percent graduating college will go up by 0.654.

(b) If the percent to graduate high school in a state is 80%, the predicted percent to graduate college is $\widehat{College} = -25.4 + 0.654 \cdot (80) = 26.92$. If the percent to graduate high school is 90%, the predicted percent to graduate college is $\widehat{College} = -25.4 + 0.654 \cdot (90) = 33.46$.

(c) The predicted college percent for Massachusetts is $\widehat{College} = -25.4 + 0.654 \cdot (86.9) = 31.43$, so the residual is $43.2 - 31.43 = 11.77$. This is the largest residual.

A.49 (a) A scatterplot of *FGPct* vs *FTPct* is shown below on the left. There is somewhat of a linear trend, and it is negative, which means that players who are better at free throws tend to be *worse* at field goals. This is an interesting result!

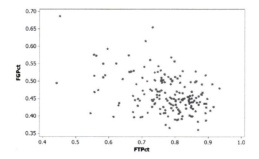
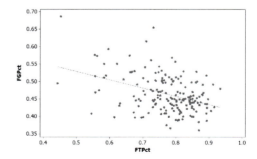

(b) There appear to be two outliers, both on the left side of the graph with free throw percentages below 50%. Both are very bad free throw shooters! One is in the middle of the pack (about 50%) at field goal percentage (this is Andrew Bogut) and the other is the best in the league at field goal percentage (this is DeAndre Jordan at 68.6% for field goals.)

(c) We use technology to find that the correlation is -0.402.

(d) The regression line is added to the plot above on the right. The formula for the regression line is $\widehat{FGPct} = 0.647 - 0.236 \cdot FTPct$.

(e) For a player with a $FTPct = 0.70$, the predicted field goal percentage is $\widehat{FGPct} = 0.647 - 0.236 \cdot (0.70) = 0.482$.

Section 3.1 Solutions

3.1 This mean is a population parameter; notation is μ.

3.3 This proportion is a sample statistic; notation is $\hat{p}$.

3.5 This mean is a sample statistic; notation is $\overline{x}$.

3.7 This is a population parameter for a mean, so the correct notation is μ. We have $\mu = 30,795/95 = 324.2$ students as the average enrollment per charter school.

3.9 This is a sample statistic from a sample of size $n = 200$ for a correlation, so the correct notation is r. We have $r = 0.037$.

3.11 This is a population parameter for a correlation, so the correct notation is ρ. We use technology to see that $\rho = -0.131$.

3.13 We expect the sampling distribution to be centered at the value of the population mean, so we estimate that the population parameter is $\mu = 85$. The standard error is the standard deviation of the distribution of sample means. The middle of 95% of the distribution goes from about 45 to 125, about 40 on either side of $\mu = 85$. By the 95% rule, we estimate that $SE \approx 40/2 = 20$. (Answers may vary slightly.)

3.15 We expect the sampling distribution to be centered at the value of the population proportion, so we estimate that the population parameter is $p = 0.80$. The standard error is the standard deviation of the distribution of sample proportions. The middle of 95% of the distribution goes from about 0.74 to 0.86, about 0.06 on either side of $p = 0.80$. By the 95% rule, we estimate that $SE \approx 0.06/2 = 0.03$. (Answers may vary slightly.)

3.17   (a) We see in the sampling distribution that a sample mean of $\overline{x} = 70$ is not unusual for samples of this size, so this value is (i): reasonably likely to occur.

(b) We see in the sampling distribution that a sample mean of $\overline{x} = 100$ is not unusual for samples of this size, so this value is (i): reasonably likely to occur.

(c) We see in the sampling distribution that a sample mean of $\overline{x} = 140$ is rare for a sample of this size but similar sample means occurred several times in this sampling distribution. This value is (ii): unusual but might occur occasionally.

3.19   (a) We see in the sampling distribution that a sample proportion of $\hat{p} = 0.72$ is rare for a sample of this size but similar sample proportions occurred several times in this sampling distribution. This value is (ii): unusual but might occur occasionally.

(b) We see in the sampling distribution that a sample proportion of $\hat{p} = 0.88$ is rare for a sample of this size but similar sample proportions occurred several times in this sampling distribution. This value is (ii): unusual but might occur occasionally.

(c) We see in the sampling distribution that there are no sample proportions even close to $\hat{p} = 0.95$ so this sample proportion is (iii): extremely unlikely to ever occur using samples of this size.

3.21 We are estimating p, the proportion of all US adults who own a laptop computer. The quantity that gives the best estimate is $\hat{p}$, the proportion of our sample who own a laptop computer. The best estimate is $\hat{p} = 1238/2252 = 0.55$. Since the true proportion is unknown, our best estimate for the proportion comes from our sample. We estimate that 55% of all US adults own a laptop computer.

3.23   (a) The value 30 is a population parameter and the notation is $\mu = 30$. The value 27.90 is a sample statistic and the notation is $\overline{x} = 27.90$.

  (b) The distribution will be bell-shaped and the center will be at the population mean of 30. The sample mean 27.90 would represent one point on the dotplot.

  (c) The dotplot will have 1000 dots and each dot will represent the mean for a sample of 75 co-payments.

3.25   (a) As the sample size goes up, the accuracy improves, which means the spread goes down. We see that distribution A goes with sample size $n = 20$, distribution B goes with $n = 100$, and distribution C goes with $n = 500$.

  (b) We see in dotplot A that quite a few of the sample proportions (when $n = 20$) are less than 0.25 or greater than 0.45, so being off by more than 0.10 would not be too surprising. While it is possible to be that far away in dotplot B (when $n = 100$), such points are much more rare, so it would be somewhat surprising for a sample of size $n = 100$ to miss by that much. None of the points in dotplot C are more than 0.10 away from $p = 0.35$, so it would be *extremely* unlikely to be that far off when $n = 500$.

  (c) Many of the points in dotplot A fall outside of the interval from 0.30 to 0.40, so it is not at all surprising for a sample proportion based on $n = 20$ to be more than 0.05 from the population proportion. Even dotplot B has quite a few values below 0.30 or above 0.40, so being off by more than 0.05 when $n = 100$ is not too surprising. Such points are rare, but not impossible in dotplot C, so a sample of size $n = 500$ might possibly give an estimate that is off by more than 0.05, but it would be pretty surprising.

  (d) As the sample size goes up, the accuracy of the estimate tends to increase.

3.27 The quantity we are trying to estimate is $p_a - p_t$ where p_a represents the proportion of adult cell phone users who text message and p_t represents the proportion of teen cell phone users who text message. The quantity that gives the best estimate is $\hat{p}_a - \hat{p}_t$, where $\hat{p}_a$ represents the proportion of the adult cell phone users in the sample of 2,252 who text message and $\hat{p}_t$ represents the proportion of teen cell phone users in the sample of 800 who text message. The best estimate for the difference in the proportion who text is $\hat{p}_a - \hat{p}_t = .72 - .87 = -0.15$.

3.29   (a) Both distributions are centered at the population parameter, so 0.05.

  (b) The proportions for samples of size $n = 100$ go from about 0 to 0.12. The proportions for samples of size $n = 1000$ go from about 0.025 to 0.07.

  (c) The standard error for samples of size $n = 100$ is about 0.02 (since it appears that about 95% of the data are between 0.01 and 0.09.) The standard error for samples of size $n = 1000$ is about 0.005 (since it appears that about 95% of the data are between 0.04 and 0.06.)

  (d) A sample proportion of 0.08 is relatively likely from a sample of 100, but extremely unlikely with a sample size of 1,000.

3.31   (a) Answers will vary. Here is one possible set of randomly selected *Points* values.
      Points: 26, 18, 3, 16, 57 $\overline{x} = 24.0$

  (b) Answers will vary. Here is another possible set of randomly selected *Points* values.
      Points: 48, 34, 13, 18, 26 $\overline{x} = 27.8$

  (c) The mean number of points for all 24 players is $\mu = 26.46$ points for the season. Most sample means found in parts (a) and (b) will be somewhat close to this but not exactly the same.

  (d) The distribution will be roughly symmetric with a peak at the center of 26.46. See the figure.

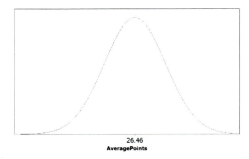

26.46
AveragePoints

3.33 Answers will vary, but a typical distribution is shown below. The smallest mean is just below 10 and the largest is just below 50 (but answers will vary). The standard deviation of these 1000 sample means is about 7.2.

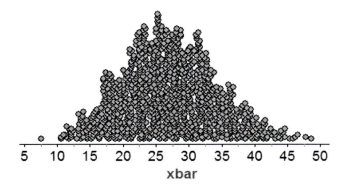

3.35 (a) This is a population proportion so the correct notation is p. We have $p = 41/273 = 0.150$.

(b) We expect it to be symmetric and bell-shaped and centered at the population proportion of 0.150.

3.37 (a) The standard error is the standard deviation of the sampling distribution (given in the upper right corner of the sampling distribution box of *StatKey*) and is likely to be about 0.11. Answers will vary, but the sample proportions should go from 0 to about 0.5 (as in the cotplot below). In that case, the farthest sample proportion from $p = 0.15$ is $\hat{p} \approx 0.5$, and it is $0.5 - 0.15 = 0.35$ off from the correct population value. In other simulations the maximum proportion might be as high as 0.6 or even 0.7.

(b) The standard error is the standard deviation of the sampling distribution and is likely to be about 0.08. Answers will vary, but the sample proportions should go from 0 to about 0.4 (as shown in the dotplot below). In that case, the farthest sample proportion from $p = 0.15$ is $\hat{p} \approx 0.4$, and it is $0.4 - 0.15 = 0.25$ off from the correct population value. Some simulations might produce even larger discrepancies.

(c) The standard error is the standard deviation of the sampling distribution and is likely to be about 0.05. Answers will vary, but the sample proportions should go from near 0 to about 0.3 (as shown in the dotplot below). In that case, the farthest sample proportion from $p = 0.15$ is $\hat{p} \approx 0.3$, and it is $0.5 - 0.15 = 0.15$ off from the correct population value. Some simulations might have even larger discrepancies.

(d) Accuracy improves as the sample size increases. The standard error gets smaller, the range of values gets smaller, and values tend to be closer to the population value of $p = 0.150$.

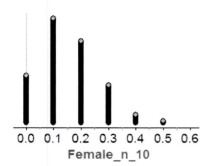

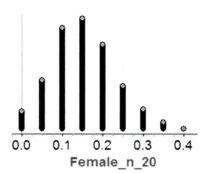

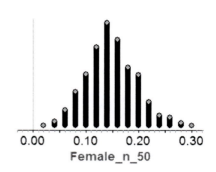

Section 3.2 Solutions

3.39 Using ME to represent the margin of error, an interval estimate for μ is $\bar{x} \pm ME = 25 \pm 3$ so an interval estimate of plausible values for the population mean μ is 22 to 28.

3.41 Using ME to represent the margin of error, an interval estimate for ρ is $r \pm ME = 0.62 \pm 0.05$ so an interval estimate of plausible values for the population correlation ρ is 0.57 to 0.67.

3.43   (a) Yes, plausible values of μ are values in the interval.

 (b) Yes, plausible values of μ are values in the interval.

 (c) No. Since 105.3 is not in the interval estimate, it is a *possible* value of μ but is not a very plausible one.

3.45 The 95% confidence interval estimate is $\hat{p} \pm 2 \cdot SE = 0.32 \pm 2(0.04) = 0.32 \pm 0.08$, so the interval is 0.24 to 0.40. We are 95% confident that the true value of the population proportion p is between 0.24 and 0.40.

3.47 The 95% confidence interval estimate is $r \pm 2 \cdot SE = 0.34 \pm 2(0.02) = 0.34 \pm 0.04$, so the interval is 0.30 to 0.38. We are 95% confident that the true value of the population correlation ρ is between 0.30 and 0.38.

3.49 The 95% confidence interval estimate is $(\bar{x}_1 - \bar{x}_2) \pm$ margin of error $= 3.0 \pm 1.2$, so the interval is 1.8 to 4.2. We are 95% confident that the true difference in the population means $\mu_1 - \mu_2$ is between 1.8 and 4.2 (which means we believe that the mean of population 1 is between 1.8 and 4.2 units larger than the mean of population 2.)

3.51   (a) The information is from a sample, so it is a statistic. It is a proportion, so the correct notation is $\hat{p} = 0.30$.

 (b) The parameter we are estimating is the proportion, p, of *all* young people in the US who have been arrested by the age of 23. Using the information in the sample, we estimate that $p \approx 0.30$.

 (c) If the margin of error is 0.01, the interval estimate is 0.30 ± 0.01 which gives 0.29 to 0.31. Plausible values for the proportion p range from 0.29 to 0.31.

 (d) Since the plausible values for the true proportion are those between 0.29 and 0.31, it is very unlikely that the actual proportion is less than 0.25.

3.53 We are 95% confident that the proportion of all adults in the US who think a car is a necessity is between 0.83 and 0.89.

3.55 We are estimating p, the proportion of all US adults who agree with the statement that each person has one true love. The best point estimate is $\hat{p} = 735/2625 = 0.28$. We find the confidence interval using:

$$
\begin{array}{ccc}
\hat{p} & \pm & 2 \cdot SE \\
0.28 & \pm & 2(0.009) \\
0.28 & \pm & 0.018 \\
0.262 & \text{to} & 0.298.
\end{array}
$$

The margin of error for our estimate is 0.018 or 1.8%. We are 95% sure that the proportion of all US adults who agree with the statement on one true love is between 0.262 and 0.298.

3.57 (a) We are 95% confident that the mean response time for game players minus the mean response time for non-players is between -1.8 to -1.2. In other words, mean response time for game players is less than the mean response time for non-players by between 1.8 and 1.2 seconds.

(b) It is not likely that they are basically the same, since the option of the difference in means being zero is not in the interval. The game players are faster, and we can tell this because the confidence interval for $\mu_g - \mu_{ng}$ has only negative values so the mean time is smaller for the game players.

(c) We are 95% confident that the mean accuracy score for game players minus the mean accuracy score for non-players is between -4.2 to 5.8.

(d) It is likely that they are basically the same, since the option of the difference in means being zero is in the interval. There is little discernible difference in accuracy between game players and non-game players.

3.59 (a) Using the margin of error, we see that the likely proportion voting for Candidate A ranges from 49% to 59%. Since this interval includes some proportions below 50% as plausible values for the election proportion, we cannot be very confident in the outcome.

(b) Using the margin of error, we see that the likely proportion voting for Candidate A ranges from 51% to 53%. Since all values in this interval are over 50%, we can be relatively confident that Candidate A will win.

(c) Using the margin of error, we see that the likely proportion voting for Candidate A ranges from 51% to 55%. Since all values in this range are over 50%, we can be relatively confident that Candidate A will win.

(d) Using the margin of error, we see that the likely proportion voting for Candidate A ranges from 48% to 68%. Since this interval includes some proportions below 50% as plausible vaues for the election proportion, we cannot be very confident in the outcome.

3.61 Let μ represent the mean time for a golden shiner fish to find the yellow mark. A 95% confidence interval is given by

$$
\begin{aligned}
\overline{x} \quad &\pm \quad 2 \cdot SE \\
51 \quad &\pm \quad 2(2.4) \\
51 \quad &\pm \quad 4.8 \\
46.2 \quad &\text{to} \quad 55.8.
\end{aligned}
$$

A 95% confidence interval for the mean time for fish to find the mark is between 46.2 and 55.8 seconds. We are 95% sure that the *mean* time it would take fish to find the target for *all* fish of this breed is between 46.2 seconds and 55.8 seconds. In other words, the plausible values for the population mean μ are those values between 46.2 and 55.8. Therefore, 60 is not a plausible value for the mean time for all fish, but 55 is.

3.63 We are estimating the difference in population proportions $p_1 - p_2$ where p_1 is the proportion of times a school of fish will pick the majority option if there is an opinionated minority, a less passionate majority, and also some additional members with no preference and p_2 is the proportion of times a school of fish will pick the majority option if there is an opinionated minority and a less passionate majority and no other fish in the group, as described above in *Fish Democracies*. (We could also have defined the proportions in the other order.) The best point estimate is $\hat{p}_1 - \hat{p}_2 = 0.61 - 0.17 = 0.44$. We find a 95% confidence interval as follows:

$$
\begin{aligned}
(\hat{p}_1 - \hat{p}_2) &\pm 2 \cdot SE \\
(0.61 - 0.17) &\pm 2(0.14) \\
0.44 &\pm 0.28 \\
0.16 \quad \text{to} &\quad 0.72.
\end{aligned}
$$

We are 95% sure that the proportion of schools of fish picking the majority option is 0.16 to 0.72 higher if fish with no preference are added to the group. If adding the indifferent fish had no effect, then the population proportions with and without the indifferent fish would be the same, which means the difference in proportions would be zero. Since zero is not a plausible value for the difference in proportions, it is very unlikely that adding indifferent fish has no effect. The indifferent fish are helping the majority carry the day.

Section 3.3 Solutions

3.65 (a) No. The value 12 is not in the original.

(b) No. A bootstrap sample has the same sample size as the original sample.

(c) Yes.

(d) No. A bootstrap sample has the same sample size as the original sample.

(e) Yes.

3.67 The distribution appears to be centered near 0.7 so the point estimate is about 0.7. Using the 95% rule, we estimate that the standard error is about 0.1 (since about 95% of the values appear to be within 0.2 of the center). Thus our interval estimate is

$$
\begin{aligned}
Statistic & \pm & 2 \cdot SE \\
0.7 & \pm & 2(0.1) \\
0.7 & \pm & 0.2 \\
0.5 & \text{to} & 0.9.
\end{aligned}
$$

The parameter being estimated is a proportion p, and the interval 0.5 to 0.9 gives plausible values for the population proportion p. Answers may vary.

3.69 The distribution appears to be centered near 0.4 so the point estimate is about 0.4. Using the 95% rule, we estimate that the standard error is about 0.05 (since about 95% of the values appear to be within 0.1 of the center). Thus our interval estimate is

$$
\begin{aligned}
Statistic & \pm & 2 \cdot SE \\
0.4 & \pm & 2(0.05) \\
0.4 & \pm & 0.1 \\
0.3 & \text{to} & 0.5.
\end{aligned}
$$

The parameter being estimated is a correlation ρ, and the interval 0.3 to 0.5 gives plausible values for the population correlation ρ. Answers may vary.

3.71 The statistic for the sample is $\hat{p} = 35/100 = 0.35$. Using technology, the standard deviation of the sample proportions for 1000 bootstrap samples is about 0.048 (answers may vary slightly), so we estimate the standard error is SE≈ 0.048. Thus our interval estimate is

$$
\begin{aligned}
Statistic & \pm & 2 \cdot SE \\
0.35 & \pm & 2(0.048) \\
0.35 & \pm & 0.096 \\
0.254 & \text{to} & 0.446.
\end{aligned}
$$

Plausible values of the population proportion range from 0.254 to 0.446.

3.73 The statistic for the sample is $\hat{p} = 112/400 = 0.28$. Using technology, the standard deviation of the sample proportions for 1000 bootstrap samples is about 0.022 (answers may vary slightly), so we estimate the standard error is SE≈ 0.022. Thus our interval estimate is

$$
\begin{array}{rcl}
Statistic & \pm & 2 \cdot SE \\
0.28 & \pm & 2(0.022) \\
0.28 & \pm & 0.044 \\
0.236 & \text{to} & 0.324.
\end{array}
$$

Plausible values of the population proportion range from 0.236 to 0.324.

3.75 (a) The best point estimate is the sample proportion, $\hat{p} = 26/174 = 0.149$.

(b) We can estimate the standard error using the 95% rule, or we can find the standard deviation of the bootstrap statistics in the upper right of the figure. We see that the standard error is about 0.028. Answers will vary slightly with other simulations.

(c) We have

$$
\begin{array}{rcl}
\hat{p} & \pm & 2 \cdot SE \\
0.149 & \pm & 2(0.028) \\
0.149 & \pm & 0.056 \\
0.093 & \text{to} & 0.205.
\end{array}
$$

We are 95% confident that the percent of all snails of this kind that will live after being eaten by a bird is between 9.3% and 20.5%.

(d) Yes, 20% is within the range of plausible values in the 95% confidence interval.

3.77 (a) The mean is $\overline{x} = 67.59$ and the standard deviation is $s = 50.02$.

(b) Select 20 values at random (with replacement) from the original set of skateboard prices and record the mean for those 20 values as the bootstrap statistic.

(c) We expect the bootstrap distribution to be symmetric and bell-shaped and to be centered at the sample mean: 67.59.

(d) We find the 95% confidence interval:

$$
\begin{array}{rcl}
\overline{x} & \pm & 2 \cdot SE \\
67.59 & \pm & 2(10.9) \\
67.59 & \pm & 21.8 \\
45.79 & \text{to} & 89.39.
\end{array}
$$

We are 95% confident that the mean price of skateboards for sale online is between $45.79 and $89.39.

3.79 (a) The best point estimate for the proportion, p, of rats showing empathy is $\hat{p} = 23/30 = 0.767$.

(b) On 23 of the slips, we write "yes" (showed empathy) and on the other 7, we write "no". We then mix up the slips of paper, draw one out and record the result, yes or no. Put the slip of paper back and repeat the process 30 times. This set of yes's and no's is our bootstrap sample. The proportion of yes's in the sample is our bootstrap statistic.

(c) Using technology, we see that the bootstrap distribution is bell-shaped and centered approximately at 0.767. We also see that the standard error is about 0.077.

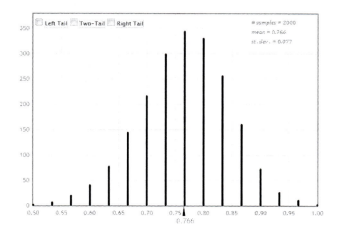

(d) We have

$$\hat{p} \quad \pm \quad 2 \cdot SE$$
$$0.767 \quad \pm \quad 2(0.077)$$
$$0.767 \quad \pm \quad 0.154$$
$$0.613 \quad \text{to} \quad 0.921.$$

For all laboratory rats, we are 95% confident that the proportion of rats that will show empathy in this manner is between 61.3% and 92.1%.

3.81 (a) The standard error is about 0.015 since most (roughly 95%) of the bootstrap distribution is between 0.12 and 0.18, which is about two standard deviations on either side of the center at 0.15.

(b) The 95% confidence interval is given by:

$$(\hat{p}_t - \hat{p}_a) \quad \pm \quad 2 \cdot SE$$
$$(0.87 - 0.72) \quad \pm \quad 2 \cdot (0.015)$$
$$.15 \quad \pm \quad 0.03$$
$$0.12 \quad \text{to} \quad 0.18.$$

We are 95% sure that the proportion of teens who text is between 0.12 and 0.18 higher than the proportion of adults who text.

3.83 (a) We are estimating μ_D, the mean difference in delay time for public transportation for all traffic situations in Dresden, Germany.

(b) Put all 24 slips in a container. Pull out one and write down the value and put it back in the container. Mix up the slips, pull out one and repeat that process until there are 24 values written down. Those 24 values form one bootstrap sample.

(c) Record the sample mean for the 24 values in the bootstrap sample.

(d) The distribution will be bell-shaped and centered at 61.

(e) We calculate the standard deviation of the bootstrap statistics.

(f) For a 95% confidence interval, we have

$$
\begin{array}{rcl}
\overline{x}_D & \pm & 2 \cdot SE \\
61 & \pm & 2(3.1) \\
61 & \pm & 6.2 \\
54.8 & \text{to} & 67.2.
\end{array}
$$

We are 95% confident that the average time savings is between 54.8 and 67.2 seconds, if the city moves to the new system.

3.85   (a) We use technology to compute the correlation between commute distances and times, $r = 0.807$, for the 500 data values.

(b) The distribution of bootstrap correlations (shown below) is fairly bell-shaped (perhaps a slight left skew), centered around 0.81, and ranges between about .70 and .90.

(c) The standard deviation of the bootstrap correlations for this bootstrap distribution is 0.0355 so the margin of error is $2 \cdot 0.0355 = 0.071$. The interval estimate for the correlation between commute distances and time is 0.807 ± 0.071 or between 0.736 and 0.878.

(d) The interval is shown on a dotplot of the bootstrap distribution below. The interval includes roughly 95% of the bootstrap correlations.

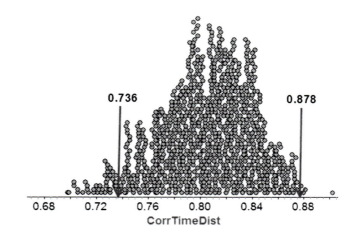

3.87 The standard deviation for the sample of penalty minutes for n=24 players is $s = 49.1$ minutes. For one set of 3000 bootstrap sample standard deviations (shown below), the estimated standard error is $SE = 11.0$.

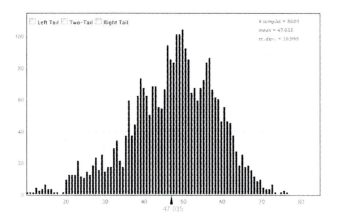

Based on this the interval estimate is

$$
\begin{array}{rcl}
s & \pm & 2 \cdot SE \\
49.1 & \pm & 2 \cdot 11.0 \\
49.1 & \pm & 22.0 \\
27.1 & \text{to} & 71.1.
\end{array}
$$

We estimate that the standard deviation in penalty minutes for all NHL players is somewhere between 27.1 and 71.1 minutes.

Section 3.4 Solutions

3.89   (a) We keep the middle 95% of values by chopping off 2.5% from each tail. Since 2.5% of 1000 is 25, we eliminate the 25 highest and the 25 lowest values to create the 95% confidence interval.

   (b) We keep the middle 90% of values by chopping off 5% from each tail. Since 5% of 1000 is 50, we eliminate the 50 highest and the 50 lowest values to create the 90% confidence interval.

   (c) We keep the middle 98% of values by chopping off 1% from each tail. Since 1% of 1000 is 10, we eliminate the 10 highest and the 10 lowest values to create the 98% confidence interval.

   (d) We keep the middle 99% of values by chopping off 0.5% from each tail. Since 0.5% of 1000 is 5, we eliminate the 5 highest and the 5 lowest values to create the 99% confidence interval.

3.91 To find a 90% confidence interval, we go less far out on either side than for a 95% confidence interval, so (C) is the most likely result.

3.93 If the sample size is smaller, we have less accuracy and the spread of the bootstrap distribution increases, so the confidence interval will be wider. Thus, (A) is the most likely result.

3.95 As long as the number of bootstrap samples is reasonable, the width of the confidence interval does not change much as we take more or fewer bootstrap samples. Thus, (B) is the most likely result.

3.97 The sample proportion who agree is $\hat{p} = 180/250 = 0.72$. One set of 1000 bootstrap proportions is shown in the figure below. For a 95% confidence interval we need to find the 2.5%-tile and 97.5%-tile, leaving 95% of the distribution in the middle. For this distribution those points are at 0.664 and 0.776, so we are 95% sure that the proportion in the population who agree is between 0.664 and 0.776. Answers will vary slightly for different simulations.

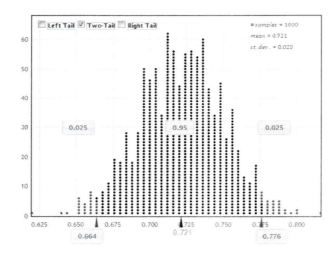

3.99 The sample proportion who agree is $\hat{p} = 382/1000 = 0.382$. One set of 1000 bootstrap proportions is shown in the figure below. For a 99% confidence interval we need to find the 0.5%-tile and 99.5%-tile, leaving 99% of the distribution in the middle. For this distribution those points are at 0.343 and 0.423, so we are 99% sure that the proportion in the population who agree is between 0.343 and 0.423. Answers will vary slightly for different simulations.

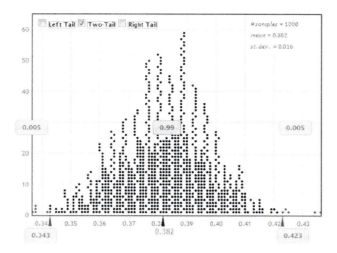

3.101 The 98% confidence interval uses the 1%-tile and 99%-tile from the bootstrap means. We are 98% sure that the mean number of penalty minutes for NHL players in a season is between 29.4 and 76.7 minutes.

3.103 Using *StatKey* or other technology, we produce a bootstrap distribution such as the figure shown below. For a 99% confidence interval, we find the 0.5%-tile and 99.5%-tile points in this distribution to be 0.467 and 0.493. We are 99% confident that the percent of all Europeans (from these nine countries) who can identify arm or shoulder pain as a symptom of a heart attack is between 46.7% and 49.3%. Since every value in this interval is below 50%, we can be 99% confident that the proportion is less than half.

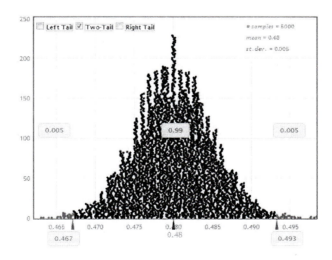

3.105 Using one bootstrap distribution (as shown below), the standard error is $SE = 0.19$.

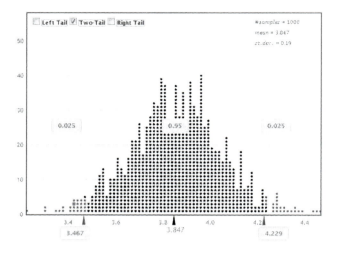

The mean tip from the original sample is $\overline{x} = 3.85$, so a 95% confidence interval using the standard error is

$$
\begin{array}{rcl}
\overline{x} & \pm & 2 \cdot SE \\
3.85 & \pm & 2(0.19) \\
3.85 & \pm & 0.38 \\
3.47 & \text{to} & 4.23.
\end{array}
$$

For this bootstrap distribution, the 95% confidence interval using the 2.5%-tile and 97.5%-tile is 3.47 to 4.23. We see that the results (rounding to two decimal places) are the same. We are 95% confident that the average tip left at this restaurant is between \$3.47 and \$4.23.

3.107 (a) We have $\hat{p}_m = 27/193 = 0.140$ and $\hat{p}_f = 16/169 = 0.095$ so the best point estimate for the difference in population proportions is $\hat{p}_m - \hat{p}_f = 0.140 - 0.095 = 0.045$. In this sample, a larger proportion of males smoke.

   (b) Using *StatKey* or other technology, we create a bootstrap distribution and find the boundaries for the middle 99% of values. We see that a 99% confidence interval for $p_m - p_f$ is the interval from about -0.039 to 0.132. We are 99% confidence that the difference between males and females in the proportion that smoke is between -0.039 and 0.132.

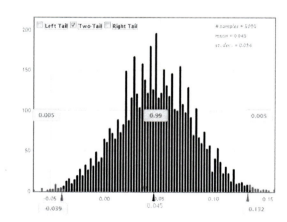

3.109   (a) We have $\overline{x}_t - \overline{x}_c = 34.82 - 17.7 = 17.12$, where $\overline{x}_t$ represents the sample mean immune response for tea drinkers and $\overline{x}_c$ represents the sample mean immune response for coffee drinkers.

(b) We are estimating $\mu_t - \mu_c$ where μ_t represents the mean immune response for all tea drinkers and μ_c represents the mean immune response for all coffee drinkers.

(c) Using *StatKey* or other technology, we obtain a bootstrap distribution of sample differences in means as shown below. We see that a 90% confidence interval for the difference in means is about 4.17 to 29.70. We are 90% confident that tea drinkers have a mean immune response between 4.17 and 29.70 higher than the mean immune response for coffee drinkers. Answers may vary for other sets of bootstrap differences in means.

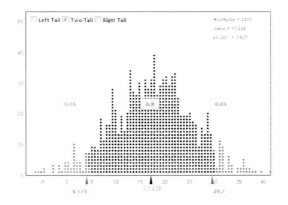 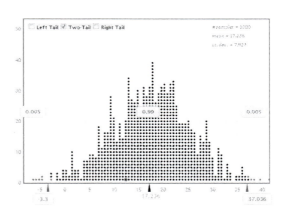

(d) Using the same bootstrap distribution, we see that a 99% confidence interval for the difference in means is about -3.30 to 37.04. We are 99% confident that the difference in mean immune response is between -3.30 and 37.04.

(e) We are 90% confident that tea drinkers have a stronger mean immune response, since all values in the 90% confidence interval are positive, but we are not 99% confident, since some plausible values for the difference in means in that interval are negative.

3.111 The mean area for the sample of ten countries is $\overline{x} = 111.3$ thousand square kilometers. Using technology we obtain a bootstrap distribution as shown below. From this distribution the 99% confidence interval is (30.1, 228.3). (Answers will vary.) We are 99% confident that the average country size for all 213 countries is between 30,100 and 228,300 square kilometers.

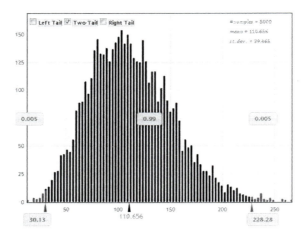

3.113   (a) We see that both cities have a significant number of outliers, with very long commute times. The quartiles and median are all bigger for Atlanta than for St. Louis, so we expect that the mean commute time is larger for Atlanta.

   (b) We are estimating the difference between the cities in mean commute time for all commuters, $\mu_{atl} - \mu_{stl}$. We get a point estimate for the difference in mean commute times between the two cities with the difference in the sample means, $\overline{x}_{atl} - \overline{x}_{stl} = 29.11 - 21.97 = 7.14$ minutes.

   (c) Since the two samples were taken independently in different cities, for each bootstrap statistic we take 500 Atlanta times with replacement from the original Atlanta data and 500 St. Louis times with replacement from the original St. Louis sample, compute the mean within each sample, and take the difference. This constitutes one bootstrap statistic.

   (d) A bootstrap distribution for the difference in means with 2000 bootstrap samples is shown in the figure.

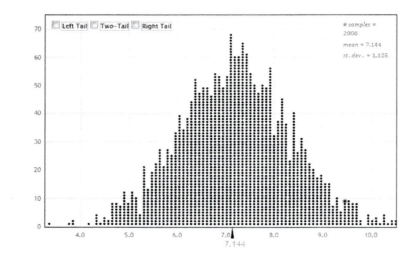

The standard error for $\overline{x}_{atl} - \overline{x}_{stl}$, found in the upper corner of the figure, is $SE = 1.125$. We find an interval estimate for the difference in the population means with

$$7.14 \pm 2 \cdot 1.125 = 7.14 \pm 2.25 = (4.89, 9.39)$$

We are 95% confident that the average commuting time for commuters in Atlanta is somewhere between 4.89 and 9.39 minutes more than the average commuting time for commuters in St. Louis.

3.115 (a) We see that the bootstrap distribution is relatively symmetric and bell-shaped, so it is reasonable to use the distribution to estimate a 95% confidence interval for the standard deviation of prices of all used Mustang cars. Using either the standard error method or the percentile method (estimating values that include the middle 95%), we estimate a 95% confidence interval to be about 7 to 14. We are 95% confident that the standard deviation of all prices of used Mustangs is between 7 thousand dollars and 14 thousand dollars.

(b) This bootstrap distribution is not symmetric and is not bell-shaped. It would not be appropriate to use this distribution to find a 95% confidence interval. The sample size is so small (at only $n = 5$) that the distribution ends up looking a bit bizarre. It is important to always look at the graph of the distribution. These methods apply only when the bootstrap distribution is reasonably symmetric and bell-shaped.

Section 4.1 Solutions

4.1   (a) We see that Sample A has the largest mean (around 30) and only one data point below 25, so it provides the most evidence for the claim that the mean placement exam score is greater than 25.

  (b) Sample C has a mean that is clearly below 25, so it provides no evidence for the claim.

4.3   (a) Sample A shows a negative association and a stronger association than Sample D, so Sample A provides the most evidence for the claim that the correlation between exam grades and time spent playing video games is negative.

  (b) In both samples B and C the the association in the scatterplots is positive, so both give no evidence for a negative correlation.

4.5 The hypotheses are:

$$H_0: \quad \mu_A = \mu_B$$
$$H_a: \quad \mu_A \neq \mu_B$$

4.7 The hypotheses are:

$$H_0: \quad \mu = 50$$
$$H_a: \quad \mu < 50$$

4.9 We define p_m to be the proportion of males who smoke and p_f to be the proportion of females who smoke. The hypotheses are:

$$H_0: \quad p_m = p_f$$
$$H_a: \quad p_m > p_f$$

4.11 We define p to be the proportion of a population who watch the Home Shopping Network. The hypotheses are:

$$H_0: \quad p = 0.20$$
$$H_a: \quad p < 0.20$$

4.13 We define μ_f to be mean study time for first year students and μ_u to be mean study time for upperclass students. The hypotheses are:

$$H_0: \quad \mu_f = \mu_u$$
$$H_a: \quad \mu_f \neq \mu_u$$

4.15   (a) These hypotheses are valid.

  (b) These hypotheses are not valid, since statistical hypotheses are statements about a population parameter (p), not a sample statistic ($\hat{p}$).

  (c) These hypotheses are not valid, since the equality should be in H_0.

  (d) These hypotheses are not valid, since a proportion, p, is always between 0 and 1 and can never be 25.

4.17 (a) We define μ_b to be the mean number of mosquitoes attracted after drinking beer and μ_w to be the mean number of mosquitoes attracted after drinking water. The hypotheses are:

$$H_0: \quad \mu_b = \mu_w$$
$$H_a: \quad \mu_b > \mu_w$$

(b) The sample mean number of mosquitoes attracted per participant before consumption for the beer group is $434/25 = 17.36$ and is $337/18 = 18.72$ for the water group. These sample means are slightly different, but the small difference could be attributed to random chance.

(c) The sample mean number of mosquitoes attracted per participant after consumption is $590/25 = 23.60$ for the beer group and is $345/18 = 19.17$ for the water group. This difference is larger than the difference in means before consumption. It is less likely to be due just to random chance.

(d) The mean number of mosquitoes attracted when drinking beer is higher than when drinking water.

(e) Since this was an experiment, a statistically significant difference would provide evidence that beer consumption increases mosquito attraction.

4.19 (a) We define μ_e to be mean BMP level in the brains of exercising mice and μ_s to be mean BMP level in the brains of sedentary mice. The hypotheses are:

$$H_0: \quad \mu_e = \mu_s$$
$$H_a: \quad \mu_e < \mu_s$$

(b) We define μ_e to be mean noggin level in the brains of exercising mice and μ_s to be mean noggin level in the brains of sedentary mice. The hypotheses are:

$$H_0: \quad \mu_e = \mu_s$$
$$H_a: \quad \mu_e > \mu_s$$

(c) We define ρ to be the correlation between levels of BMP and noggin in the brains of mice. The hypotheses are:

$$H_0: \quad \rho = 0$$
$$H_a: \quad \rho < 0$$

4.21 We define μ_m to be mean heart rate for males being admitted to an ICU and μ_f to be mean heart rate for females being admitted to an ICU. The hypotheses are:

$$H_0: \quad \mu_m = \mu_f$$
$$H_a: \quad \mu_m > \mu_f$$

4.23 We define ρ to be the correlation between systolic blood pressure and heart rate for patients admitted to an ICU. The hypotheses are:

$$H_0: \quad \rho = 0$$
$$H_a: \quad \rho > 0$$

Note: The hypotheses could also be written in terms of β, the slope of a regression line to predict one of these variables using the other.

4.25 We define μ to be the mean age of ICU patients. The hypotheses are:

$$H_0: \quad \mu = 50$$
$$H_a: \quad \mu > 50$$

4.27 (a) The population parameter of interest is the correlation, ρ, between number of children and household income from all households within the US. The hypotheses are:

$$H_0: \quad \rho = 0$$
$$H_a: \quad \rho \neq 0$$

   (b) We are testing for a relationship, which means a correlation different from 0. So the larger correlation of 0.75 provides more evidence.

   (c) Since we are simply looking for evidence of a relationship, the same amount of evidence is given from both 0.50 and -0.50. However they would give evidence of a relationship in opposite directions.

4.29 (a) We define μ to be the average amount of omega-3 in one tablespoon of this brand of milled flaxseed. Since the company is looking for evidence that the average is greater than 3800, the hypotheses are:

$$H_0: \quad \mu = 3800$$
$$H_a: \quad \mu > 3800$$

   (b) We define μ as in part (a). Since the consumer organization is testing to see if there is evidence that the average is less than 3800, the hypotheses are:

$$H_0: \quad \mu = 3800$$
$$H_a: \quad \mu < 3800$$

4.31 This analysis does involve a statistical test. The population parameter p is the proportion of people in the community living in a mobile home. The hypotheses are:

$$H_0: \quad p = 0.10$$
$$H_a: \quad p > 0.10$$

4.33 This analysis does include a statistical test. This is a matched pairs experiment, so the population parameter is μ_D, the average difference of reaction time (right $-$ left) for all right-handed people. The hypotheses are $H_0: \mu_D = 0$ vs $H_a: \mu_D < 0$. We could also write the hypotheses as $H_0: \mu_R = \mu_L$ vs $H_a: \mu_R < \mu_L$.

4.35 This analysis does include a statistical test for a single proportion. The population parameter p is the proportion of people in New York City who prefer Pepsi and we are testing to see if the proportion who prefer Pepsi is greater than 50%. The hypotheses are:

$$H_0: \quad p = 0.5$$
$$H_a: \quad p > 0.5$$

4.37 (a) We define p_c to be the proportion supporting Candidate A after a phone call and p_f to be the proportion supporting Candidate A after a flyer. The hypotheses are:

$$H_0: \quad p_c = p_f$$
$$H_a: \quad p_c \neq p_f$$

(b) We see that $\hat{p}_c = 152/250 = 0.608$ and $\hat{p}_f = 145/250 = 0.580$. The sample proportions are not equal.

(c) We see that $\hat{p}_c = 188/250 = 0.752$ and $\hat{p}_f = 120/250 = 0.480$.

(d) Sample B has stronger evidence of a difference in effectiveness because the sample proportions are farther apart than in Sample A and the sample sizes are the same.

4.39 (a) The parameter of interest is ρ, the correlation between score on the mouse grimace scale and pain intensity and duration. Since the study is investigating a positive relationship between these variables, the hypotheses are $H_0 : \rho = 0$ vs $H_a : \rho > 0$.

(b) Yes. If they conclude there is some relationship, the sample correlation must have been statistically significant (and positive).

(c) No. If the original correlation is statistically significant, a sample produced under the null hypothesis of no relationship should rarely give a correlation more extreme than was originally observed. This is what we mean by statistically significant.

(d) If the results of the original study were not significant, it would not be very unusual to get a sample correlation that extreme when H_0 is true. So it would not be very surprising to see a placebo give a larger correlaion.

Section 4.2 Solutions

4.41 The smaller p-value, 0.08, provides stronger evidence against H_0.

4.43 The smaller p-value, 0.007, provides stronger evidence against H_0.

4.45 This is a right-tail test, so in each case, we are estimating the proportion of the distribution to the right of the value.

(a) $\bar{x} = 68$ is far in the right tail of the randomization distribution so the p-value is quite small at 0.01.

(b) $\bar{x} = 54$ relatively close to the center of the distribution with roughly 1/3 of the values in the upper tail beyond it, so the p-value is closer to 0.30 than it is to 0.10.

(c) $\bar{x} = 63$ is well out in the upper tail with relatively few (about 5%) of the values above it so the p-value is closer to 0.05. A p-value near 0.50 would come from some $\bar{x}$ in the middle of the randomization distribution, near 50.

4.47 This is a two-tail test, so in each case, we find the area beyond the value in the smaller tail and then double it.

(a) $D = -2.9$ is far out in the lower tail of the randomization distribution. Even accounting for the two tails, there are very few points this extreme so the p-value must be quite small, around 0.01.

(b) $D = 1.2$ is in the upper tail and it's p-value uses the points in the tail beyond it, together with those beyond -1.2 in the other tail. This still leaves more than half of the points in the randomization distribution between-1.2 and +1.2, so the p-value must be less than 0.50. A p-value near 0.30 would be a good estimate of the amount of the distribution beyond ± 1.2..

4.49 (a) The figures showing the (two-tail) region beyond the observed statistic are shown below.

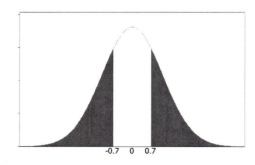 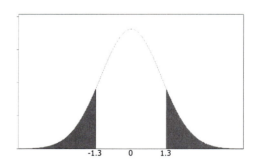

(b) There is less areas in the tails beyond ± 1.3 than beyond ± 0.7. So $D = -1.3$ has a smaller p-value and provides stronger evidence against H_0.

4.51 (a) We compute the difference in means: $D = \bar{x}_1 - \bar{x}_2 = 95.7 - 93.5 = 2.2$ and $D = \bar{x}_1 - \bar{x}_2 = 94.1 - 96.3 = -2.2$. The figures are shown below.

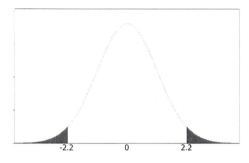
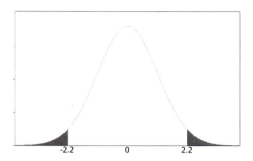

(b) The sample means for both pairs, $D = \bar{x}_1 - \bar{x}_2 = 95.7 - 93.5 = 2.2$ and $D = \bar{x}_1 - \bar{x}_2 = 94.1 - 96.3 = -2.2$ differ by the same amount, but with opposite signs. Since the p-value for this two tail test uses the values that are more extreme from *both* tails, each pair produces the same p-value. The evidence against H_0 is identical in both cases.

4.53 The test showing statistical evidence of improvement should have a small p-value, while the test showing no significant improvement should have a larger p-value. Since mice exposed to UV did "significantly better", that test should have the small p-value of 0.002. There was no significant improvement for the mice given vitamin D, so that test should have the large p-value of 0.472.

4.55 (a) This is a right-tail test so we shade the area to the right of the statistic 1.6. See the figure. The amount of the distribution that lies to the right of 1.6 is a relatively small portion of the entire graph. It is not as small as 0.03 and is not close to half the data so is not as large as 0.45 or 0.60. The p-value of 0.11 is the most reasonable estimate.

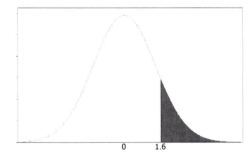
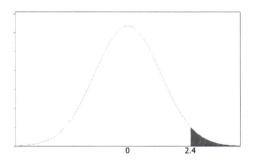

(b) See the figure. The amount of the distribution that lies to the right of the statistic $D = 2.4$ is very small, so the p-value is closest to $30/1000 = 0.03$.

(c) The difference in part (b), $D = 2.4$, is more extreme in the randomization distribution and has a smaller p-value, so it provides stronger evidence that caffeine increases average finger tapping rate.

4.57 (a) Since this is a two-tail test, we shade *both* tails, first beyond ± 0.2, then beyond ± 0.4. See the figure.

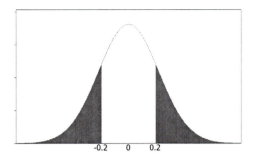 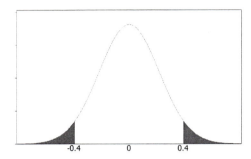

(b) Less than half the area of the distribution lies in the two tails beyond ± 0.2, but not a lot less than half, so the p-value of 0.392 would be the best estimate. From the frequencies in the histogram, there appear to be somewhere between 25 and 35 cases below $D = -0.4$, so doubling to account for two tails would make 66/1000=0.066 the most reasonable p-value of those listed.

(c) The smallest p-value provides the strongest evidence that the proportions differ and the methods (phone calls and flyers) are not equally effective. Among the five choices in part (b), the smallest p-value is 0.008.

4.59   (a) Finding significant differences on kindergarten tests would mean the p-value is relatively small.

(b) Finding no significant difference on junior high and high school tests would mean the p-value is not small and is relatively large.

(c) Finding significant differences again in adulthood indicates the p-values for testing those factors are relatively small.

4.61 The randomization distribution represents samples chosen when H_0 is true. The area in a tail gives an estimate of the probability that a result as extreme (or more extreme) than the original sample should occur when H_0 is true, which is what the p-value measures.

4.63   (a) For the small p-value of 0.0012, we expect about $0.0012 \cdot 1000 = 1.2$ or about one time out of every 1000 to be as extreme as the difference observed, if the questions are equally difficult.

(b) The p-value is very small, so seeing this large a difference would be very unusual if the two questions really were equally difficult. Thus we conclude that there is a difference in the average difficulty of the two questions.

(c) There is nothing in the information given that indicates which question had the higher mean, so we can't tell which of the two questions is the easier one.

4.65   (a) Since a fair die has six equally likely results, we should roll a five on 1/6 of all throws. The question asks if fives are more common than would be expected for a fair die, so this is a right-tail test with $H_0 : p = 1/6$ vs $H_a : p > 1/6$, where p is the proportion of throws that show a five.

(b) If $H_0 : p = 1/6$ is true, the sample proportions should be centered around the null proportion $p = 1/6$.

(c) Answers can vary, and any value of $\hat{p} < 1/6 = 0.1666...$ will work.

(d) If $\hat{p} < 1/6$ it will be to the left of the center of the randomization distribution at $p = 1/6$.

(e) Since this is an upper tail H_a, the p-value is found by the part of the randomization that is above the observed statistic, so we find the area to the right.

(f) Since a sample $\hat{p}$ below 1/6 is in the left half of the randomization distribution, the p-value above it must be more than 0.50.

Section 4.3 Solutions

4.67 Reject H_0, since p-value $= 0.0007 < 0.05$.

4.69 Do not reject H_0, since p-value $= 0.2531 \geq 0.05$.

4.71 The results are significant if the p-value is less than the significance level. A p-value of 0.0320 shows the results are significant at a 10% and 5% level, but not a 1% level.

4.73 The results are significant if the p-value is less than the significance level. A p-value of 0.008 shows the results are significant at all three levels.

4.75 The results are significant if the p-value is less than the significance level.

  (a) I. 0.0875, less than 0.10, but not less than 0.05.

  (b) IV. 0.00003, smaller than any reasonable significance level.

  (c) II. 0.5457, larger than any reasonable significance level.

  (d) III. 0.0217, less than 0.05, but not less than 0.01.

4.77 Test A, since we can be more sure the p-value is quite small. It's likely that Test B has a p-value between 0.05 and 0.10, since otherwise they probably would have reported it as being significant at a smaller level than 10%.

4.79 If we use a 5% significance level, the p-value of 0.165 is not less than $\alpha = 0.05$ so we would not reject $H_0 : p_f = p_m$. This means the data do not show significant evidence of a difference in the proportions of men and women that view divorce as "morally acceptable".

4.81 The game will last longer with the smaller significance level, 1%, that requires stronger evidence to determine a winner for competition. Any game standings that are significant at a 1% level will also be significant at a 5% level, but the results may become significant with a p-value below 5% before the p-value gets below 1%.

4.83   (a) Students who think the drink is more expensive solve, on average, more puzzles than students who have a discounted price. The p-value is very small so the evidence for this conclusion is very strong.

  (b) If you price a product too low, customers might perceive it to be less effective or lower quality than it actually is.

4.85   (a) The hypotheses are $H_0 : p = 0.5$ vs $H_a : p \neq 0.5$ where p is the proportion of penalty shots that go to the right after a specific body movement.

  (b) The p-value (0.3184) is not small so we do not reject H_0. There is not evidence that this movement helps predict the direction of the shot, so there is no reason to learn to distinguish it.

  (c) The p-value (0.0006) is smaller than any common significance level so we reject H_0. The proportion of shots to the right after this movement is different from 0.50, so a goalie can gain an advantage by learning to distinguish this movement.

4.87   (a) Since the results were statistically significant at a 5% significance level, we know the p-value was less than 0.05.

(b) Since we aren't given an exact p-value and can only deduce that the p-value is less than 0.05, we can't determine whether the evidence is very strong or only moderately strong.

(c) Yes, since the results are statistically significant we can conclude that pesticide exposure is related to ADHD.

(d) No. The data were collected with an observational study and not an experiment so we should not draw a cause and effect conclusion.

4.89 (a) Yes, since the p-value is very small, there is strong evidence against the null and in favor of the alternative that more mosquitoes, on average, are attracted to those who drink beer than those who drink water.

(b) A p-value less than 0.001 indicates that such extreme data would be very rare if beer and water had the same effect, so there is very strong evidence against the null hypothesis.

(c) Since the data were collected with an experiment where the conditions (beer or water) were randomly assigned, we can infer from the small p-value that consuming beer causes an increase in mosquito attraction.

4.91 (a) We know the two-tailed p-value is less than 0.05, but can't tell if it's also less than 0.01, so we can't make a decision for the 1% test.

(b) If we reject H_0 at a 5% level, the p-value is less than 0.05, so it is also less than 0.10 and thus we would also reject H_0 at a 10% level.

(c) If the sample correlation $r > 0$, its p-value in an upper tail test would be one half of the two-tailed p-value. Since the two-tailed p-value is less than 0.05, half of it will be even smaller and thus also less than 0.05. Since the p-value for the one-tailed test is less than 5%, H_0 will be rejected at a 5% level, so the conclusion is valid. (Note, however, if the sample $r < 0$ (i.e. is in the lower tail) then it's upper tail p-value would be more than 0.50 and show no evidence to reject $H_0 : \rho = 0$ in favor of $H_a : \rho > 0$.)

4.93 A Type I error (releasing a drug that is really not more effective) when there are serious side effects should be avoided, so it makes sense to use a small significance level such as $\alpha = 0.01$.

4.95 A Type I error (saying your average Wii bowling score is higher than a friend's, when it isn't) is not very serious, so a large significance level such as $\alpha = 0.10$ will make it easier to see any difference.

4.97 A Type I error (getting people to take the supplements when they don't help) is not serious if there are no harmful side effects, so a large significance level, such as $\alpha = 0.10$, will make it easier to see any benefit of the supplements.

4.99 Type I error: Release a drug that is really not more effective. Type II error: Fail to show the drug is more effective , when actually it is. Personal opinions will vary on which is worse.

4.101 Type I error: Find your average Wii bowling score is higher than a friend's, when actually it isn't. Type II error: The sample does not contain enough evidence to conclude that your average Wii bowling score is better than your friend's mean, when actually it is better. Personal opinions will vary on which is worse.

4.103 Type I error: Conclude people should take the supplements when they actually don't help. Type II error: Fail to detect that the supplements are beneficial, when they actually are. Personal opinions will vary on which is worse.

4.105   (a) If the sample shows significant results we reject H_0. If that conclusion is right, we have not made an error. If that conclusion is wrong (i.e., H_0 is true) we've made a Type I error.

(b) If the sample shows insignificant results we do not reject H_0. If that conclusion is right, we have not made an error. If that conclusion is wrong (i.e., H_0 is false) we've made a Type II error.

(c) We would need to know the actual value of the parameter for the population to verify if we made the correct decision or an error, and if we knew the actual value of the parameter we would not need to do any statistical inference.

Section 4.4 Solutions

4.107 The sample proportion, $\hat{p}$

4.109 The sample correlation, r

4.111 The difference in the sample proportions, $\hat{p}_1 - \hat{p}_2$

4.113 The randomization distribution will be centered at 10, since $\mu = 10$ under H_0. Since $H_a : \mu > 10$, this is a right-tail test.

4.115 The randomization distribution will be centered at 0, since $\mu_1 - \mu_2 = 0$ under H_0. Since $H_a : \mu_1 \neq \mu_2$, this is a two-tailed test.

4.117 A randomization distribution with sample proportions for 1000 samples of size 50 when $p = 0.50$ is shown below. Since the alternative hypothesis is $p > 0.5$, this is a right-tail test and we need to find how many of the randomization proportions are at (or above) the sample value of $\hat{p} = 0.60$. For this randomization distribution that includes 102 out of 1000 simulations so the p-value is 0.102. Answers will vary.

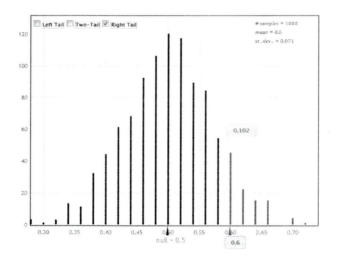

4.119 A randomization distribution with sample proportions for 1000 samples of size 200 when $p = 0.70$ is shown below. Since the alternative hypothesis is $p < 0.7$, this is a left-tail test and we need to find how many of the randomization proportions are at (or below) the sample value of $\hat{p} = 0.625$. For this randomization distribution that includes 9 out of 1000 simulations so the p-value is 0.009. Answers will vary.

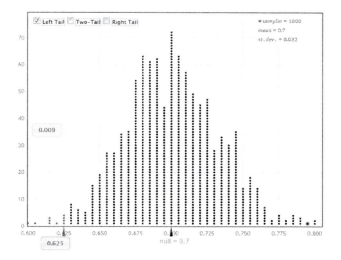

4.121 A randomization distribution with sample proportions for 1000 samples of size 100 when $p = 0.5$ is shown below. Since the alternative hypothesis is $p \neq 0.5$, this is a two-tail test. We need to find how many of the randomization proportions are at (or below) the sample value of $\hat{p} = 0.42$ and double to account for the other tail. For this randomization distribution there are 60 out of 1000 simulations beyond 0.42 so the p-value is $2 * 60/1000 = 0.120$. Answers will vary.

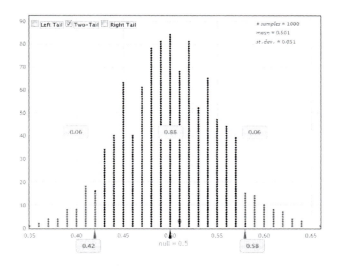

4.123 (a) The hypotheses are $H_0 : p = 0.5$ vs $H_a : p > 0.5$, where p is the proportion of all games Paul the Octopus picks correctly.

  (b) Answers vary, but 8 out of 8 heads should rarely occur.

  (c) The proportion of heads in flipping a coin is $p = 0.5$ which matches the null hypothesis.

4.125 Since you are testing whether the coin is biased, the null hypothesis is $H_0 : p = 0.5$, where p is the proportion of heads for all flips. We assume the coin is fair and H_0 is true when we create the randomization distribution, so we expect the simulated sample $\hat{p}$ values to be centered around 0.5.

4.127 (a) This is a difference in means test. If we let μ_{LL} represent mean weight gain for mice in bright light all the time and μ_{LD} represent mean weight gain for mice in a normal light/dark cycle, we have:

$$H_0: \quad \mu_{LL} = \mu_{LD}$$
$$H_a: \quad \mu_{LL} > \mu_{LD}$$

(b) We assume the null hypothesis is true to create a randomization sample, which means we assume that it is irrelevant which light situation each mouse was in. We simulate this situation by taking the 18 data values and randomly assigning them to the two conditions. One possible sample (out of many many possibilities) is shown in the following table. Notice that the values are the same as in the original table, but they have been randomly rearranged between the groups (matching the null hypothesis that says the light condition doesn't matter.)

Bright light	6	4	7	17	10	9	11	8	10
Light/Dark	11	9	6	8	12	3	5	6	9

4.129 We use technology to create randomization samples by repeatedly sampling six values with replacement (after subtracting 11 from each value in the original sample to match the null hypothesis that $\mu = 80$). We collect the means for 1000 such samples to form a randomization distribution such as the one shown below. To find the p-value for this right-tailed alternative we count how many of the randomization means exceed our original sample mean, $\overline{x} = 91.0$. For the distribution below this is 115 of the 1000 samples, so the p-value $= 0.115$. This is not enough evidence (using a 5% significance level) to reject $H_0 : \mu = 80$. The chain should continue to order chickens from this supplier (but also keep testing the arsenic level of its chickens).

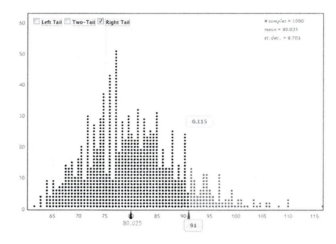

4.131 This is a test for a single proportion, since there is only one sample of 95 attacks. We can test the proportion of attacks before the full moon, or after – it doesn't matter which since if we know one we know the other. We let p represent the proportion of attacks that happen after the full moon. For the sample data we see $\hat{p} = 71/95 = 0.747$. Since "equally split" means $p = 0.5$, we have as our hypotheses:

$$H_0: \quad p = 0.5$$
$$H_a: \quad p > 0.5$$

We use *StatKey* or other technology to create a randomization distribution using proportions for samples of size $n = 95$ simulated when $p = 0.5$ to match this null hypothesis. One such randomization distribution is shown below. Since the alternative hypothesis is $H_a : p > 0.5$, this is a right-tail test. We see that the sample statistic $\hat{p} = 71/95 = 0.747$ is well beyond *any* of the randomization proportions, so we see that the p-value is essentially zero. At any significance level, we reject H_0. This data provides *very* strong evidence that lions are more likely to attack during the five days after a full moon than during the five days before. Watch out for lions after a full moon!

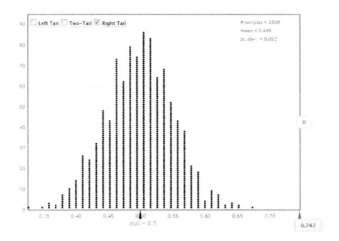

4.133 We are testing whether the correlation ρ is positive, where ρ represents the correlation between the malevolence rating of uniforms in the National Hockey League and team z-scores for the number of penalty minutes. The hypotheses for the test are

$$H_0 : \quad \rho = 0$$
$$H_a : \quad \rho > 0$$

Using the data for *NHL_Malevolence* and *ZPenMin* in the dataset **MalevolentUniformsNHL** and using either StatKey or other technology, we match the null assumption that the correlation is zero by randomly assigning the values for one of the variables to the values of the other variable and compute the correlation of the resulting simulated data. We do this 1000 times and collect these simulated correlation statistics into a randomization dotplot (shown below). To see how extreme the original sample correlation of $r = 0.521$ is we find the proportion of simulated samples that have a correlations of 0.521 or larger. For the distribution below, this includes 18 of the 1000 randomizations in the right tail, giving a p-value $18/1000 = 0.018$. Answers will vary for other randomizations. Using a 5% significance level, we reject H_0 and conclude that the malevolence of hockey uniforms is positively correlated with the number of minutes a team is penalized. The results are significant at a 5% level but not at a 1% level.

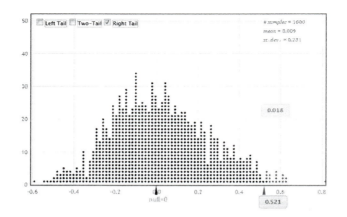

4.135 (a) The hypotheses are $H_0 : p_1 = p_2$ vs $H_a : p_1 > p_2$, where p_1 and p_2 are the proportion with reduced pain when using cannabis and a placebo, respectively.

(b) The sample statistics are $\hat{p}_1 = 14/27 = 0.519$ and $\hat{p}_2 = 7/28 = 0.250$. Since $\hat{p}_1 > \hat{p}_2$, the sample statistics are in the direction of H_a.

(c) If the FDA requires very strong evidence to reject H_0, they should choose a small significance level, such as $\alpha = 0.01$.

(d) In this situation, under a null hypothesis that $H_0 : p_1 = p_2$, pain response would be the same whether in the cannabis or placebo group. The randomization distribution for $\hat{p}_1 - \hat{p}_2$ should be centered at zero.

(e) We draw a bell-shaped curve, centered at 0, and roughly locate the original sample statistic, $\hat{p}_1 - \hat{p}_2 = 0.519 - 0.250 = 0.269$, so that the area in the right tail is only about 0.02.

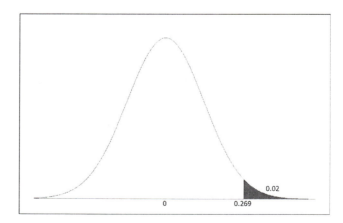

(f) A p-value as small as 0.02 gives fairly strong evidence to reject $H_0 : p_1 = p_2$.

(g) If the FDA uses a small $\alpha = 0.01$, the p-value is not less than α, so we do not reject H_0. Although the sample results are suggestive of a benefit to using cannabis for pain reduction, they are not sufficiently strong to conclude (at a 1% significance level) that the proportion of patients having reduced pain after using cannabis is more than the proportion who are helped by a placebo.

4.137 (a) It doesn't matter whether we test the proportion who plan to vote for Obama or the proportion who plan to vote for McCain, since if we know one we know the other. (For example, if 54% plan to vote for Obama, then the other 46% plan to vote for McCain and vice versa.) If we let p_O represent the proportion that plan to vote for Obama, then we are testing to see if the proportion is greater than 0.5 so the hypotheses are

$$H_0: \quad p_O = 0.5$$
$$H_a: \quad p_O > 0.5$$

Notice that if we use p_M as the proportion that plan to vote for McCain, the alternative hypothesis would be $p_M < 0.5$. The two ways of doing the test are equivalent.

(b) We assume from the null hypothesis that half ($p_O = 0.5$) plan to vote for Obama and half for McCain, so we might flip a coin to determine the outcome for each simulated voter. Since the sample size was 1057, to create a simulated sample, we need to flip a coin 1057 times (or flip 1057 coins!) and record the number of heads (Obama) and tails (McCain) in the 1057 flips. We record the proportion for Obama as one randomization statistic. We do this many times to collect 1000 or more such randomization statistics and those values form our randomization distribution. There are many other ways to create the randomization distribution, but every such method should use the null assumption of half supporting each candidate and every method should use the sample size of 1057.

4.139 (a) The hypotheses are $H_0: p_d = p_c$ vs $H_a: p_d < p_c$, where p_d and p_c are the proportion who relapse when treated with desipramine or a placebo, respectively.

(b) Under H_0 we have $p_d - p_c = 0$, so the distribution of differences in number relapsing should be centered at zero.

(c) Start with 48 cards, put "R" for relapse on 30 of them, leave the others blank (no relapse). Shuffle the cards and deal into two piles of 24 each, one for the desipramine group, the other for the placebo group. Count the number of relapses in the desipramine group and subtract the number in the placebo group.

4.141 (a) We are interested in whether pulse rates are higher on average than lecture pulse rates, so our hypotheses are

$$H_0: \quad \mu_Q = \mu_L$$
$$H_a: \quad \mu_Q > \mu_L$$

where μ_Q represents the mean pulse rate of students taking a quiz in a statistics class and μ_L represents the mean pulse rate of students sitting in a lecture in a statistics class. We could also word hypotheses in terms of the mean difference $D=Lecture\ pulse-Quiz\ pulse$ in which case the hypotheses would be $H_0: \mu_D = 0$ vs $H_a: \mu_D > 0$.

(b) We are interested in the difference between the two pulse rates, so an appropriate statistic is $\overline{x}_D$, the differences ($D = Lecture - Quiz$) for the sample. For the original sample the differences are:

$$+2, \quad -1, \quad +5, \quad -8, \quad +1, \quad +20, \quad +15, \quad -4, \quad +9, \quad -12$$

and the mean of the differences is $\overline{x}_D = 2.7$.

(c) Since the data were collected as pairs, our method of randomization needs to keep the data in pairs, so the first person is labeled with (75, 73), with a difference in pulse rate of 2. As long as we keep the data in pairs, there are many ways to conduct the randomization. In every case, of course, we need to make sure the null hypothesis (no difference) is met and we need to keep the data in pairs.

One way to do this is to sample from the pairs with replacement, but randomly determine the order of the pair (perhaps by flipping a coin), so that the first pair might be $(75, 73)$ with a difference of 2 or might be $(73, 75)$ with a difference of -2. Notice that we match the null hypothesis that the quiz/lecture situation has no effect by assuming that it doesn't matter – the two values could have come from either situation. Proceeding this way, we collect 10 differences with randomly assigned signs (positive/negative) and compute the average of these differences. That gives us one simulated statistic.

A second possible method is to focus exclusively on the differences as a single sample. Since the randomization distribution needs to assume the null hypothesis, that the mean difference is 0, we can subtract 2.7 (the mean of the original sample of differences) from each of the 10 differences, giving the values
$$-0.7, \quad -3.7, \quad 2.3, \quad -10.7, \quad -1.7, \quad 17.3, \quad 12.3, \quad -6.7, \quad 6.3, \quad -14.7.$$
Notice that these values have a mean of zero, as required by the null hypothesis. We then select samples of size ten (with replacement) from the adjusted set of differences (perhaps by putting the 10 values on cards or using technology) and compute the average difference for each sample.

There are other possible methods, but be sure to use the paired data values and be sure to force the null hypothesis to be true in the method you create!

(d) Here is one sample if we randomly assign $+/-$ signs to each difference:
$$+2, \quad +1, \quad -5, \quad -8, \quad +1, \quad -20, \quad -15, \quad +4, \quad +9, \quad +12 \qquad \Rightarrow \overline{x}_D = -1.9$$

Here is one sample drawn with replacement after shifting the differences.
$$-6.7, \quad -1.7, \quad 6.3, \quad 2.3, \quad -3.7, \quad -1.7, \quad -0.7, \quad 17.3, \quad 2.3, \quad -0.7 \qquad \Rightarrow \overline{x}_D = 1.3$$

(e) Neither of the statistics for the randomization samples in (d) exceed the value of $\overline{x}_D = 2.7$ from the original sample, but your answers will vary for other randomizations.

4.143 (a) Sampling 100 responses (with replacement) from the original sample is inappropriate, since the samples are taken from a "population" where the proportion is 0.76, so it doesn't match $H_0 : p = 0.8$.

(b) Sampling 100 responses (with replacement) from a set consisting of 8 correct responses and 2 incorrect responses is appropriate, since $p = 0.8$ in this population and we are taking the same size samples as the original data.

4.145 We are testing $H_0 : \mu_F = \mu_M$ vs $H_0 : \mu_F \neq \mu_M$, where μ represents the average hours of exercise in a week. The difference for the original sample is $\overline{x}_F - \overline{x}_M = 9.40 - 12.40 = -3.00$. We need to see how many randomizations samples give mean differences this small (or even more extreme), then double the count to account for the other tail of the randomization distribution since this is a two-tailed alternative. Note, since this is a two-tailed test, we can just as easily use $\overline{x}_M - \overline{x}_F = 12.40 - 9.40 = +3.00$ as the difference for the original sample and look for more extreme values in the other tail.

(a) We randomly scramble the gender labels (M and F) and pair them with the actual exercise times, then find the mean exercise time within each randomly assigned group. (*StatKey* tip: Use "Reallocate" as the method of randomization.) For each randomization we record the difference in means $\overline{x}_F - \overline{x}_M$. Repeating this for 1000 randomizations produces a distribution such as the one shown below. To find the p-value we count the number of differences at (or below) the original sample difference of -3.00 and double the count to account for the other tail. In this case we have p-value= $2 * 99/1000 = 0.198$.

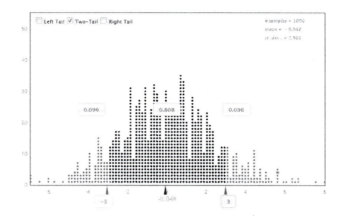

(b) Depending on technology, for this approach it may help to create two separate samples, one for the females and one for the males. Add 1.2 to each of the exercise values in the female sample and subtract 1.8 from each of the exercise values in the male sample to create new variables that have the same mean (10.6) in each group. Take separate samples (with replacement) — size 30 from the females and size 20 from the males — and find the mean exercise value in each sample. To find the randomization distribution we collect 1000 such differences. Depending on technology it might be easier to generate 1000 means for each gender and then subtract to get the 1000 differences. (*StatKey* tip: Use "Shift Groups" as the method of randomization.) One set of 1000 randomization differences in means is shown below. This distribution has 98 sample differences less than (or equal to) the original difference of -3.00, so the two-tailed p-value= $2 \cdot 98/1000 = 0.196$.

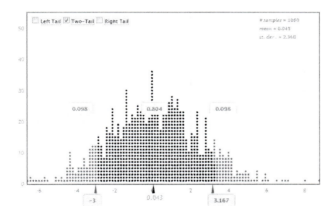

(c) We sample 30 values (with replacement) from the original sample of all 50 exercise amounts to simulate a new sample females and do the same for 20 males. Since both samples are drawn from the same set, we satisfy $H_0 : \mu_F = \mu_M$. (*StatKey* tip: Use "Combine Groups" as the method of randomization.) Compute the difference in means, $\bar{x}_F - \bar{x}_M$, for the new samples. One set of 1000 randomizations using this method is shown below. We see that 99 differences in means are at (or below) the original -3.00, so the two-tailed p-value= $2 \cdot 99/1000 = 0.198$.

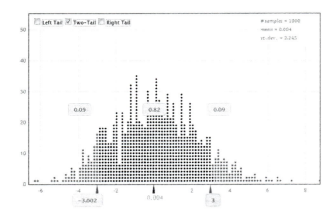

Note that that the randomization distributions and p-values produced by each method are similar. In each case the p-value is not small and we have insufficient evidence to conclude that there is difference in mean exercise time between female and male students.

Section 4.5 Solutions

4.147 (a) Since the null $p = 0.5$ is outside of the 95% confidence interval, we reject H_0 using a 5% significance level.

(b) Since the null $p = 0.5$ is in the 95% confidence interval, we do not reject H_0 using a 5% significance level.

(c) Since the null $p = 0.5$ is in the 99% confidence interval, we do not reject H_0 using a 1% significance level.

4.149 (a) Since the null $\mu_1 - \mu_2 = 0$ is outside of the 95% confidence interval, we reject H_0 using a 5% significance level. The interval includes only positive values, suggesting μ_1 is larger than μ_2.

(b) Since the null $\mu_1 - \mu_2 = 0$ is in the 99% confidence interval, we do not reject H_0 using a 1% significance level.

(c) Since the null $\mu_1 - \mu_2 = 0$ is outside of the 90% confidence interval, we reject H_0 using a 10% significance level. The interval includes only negative values, suggesting μ_2 is larger than μ_1.

4.151 (a) Since the null $\mu = 100$ is below the 99% confidence interval, we reject H_0 using a 1% significance level.

(b) Since the null $\mu = 150$ is inside the 99% confidence interval, we do not reject H_0 using a 1% significance level.

(c) Since the null $\mu = 200$ is above the 99% confidence interval, we reject H_0 using a 1% significance level.

4.153 (a) The hypotheses are $H_0 : p = 0.20$ and $H_a : p \neq 0.20$. The hypothesized proportion of 0.20 lies *inside* the confidence interval 0.167 to 0.216, so 0.20 is a reasonable possibility for the population proportion given the sample results. Using a 5% significance level, we do not reject H_0 and do not find evidence that Congressional approval is different than 20%.

(b) The hypotheses are $H_0 : p = 0.14$ and $H_a : p \neq 0.14$. The hypothesized proportion of 0.14 lies *outside* the confidence interval 0.167 to 0.216, so 0.14 is *not* a reasonable possibility for the population proportion given the sample results. Using a 5% significance level, we reject H_0 and conclude that Congressional approval in August 2010 is different than the record low of 14%.

4.155 (a) Randomly assign half the subjects to read a passage in a print book and the other half to read the same passage with an e-book. Record the time needed to complete the reading for each participant. Other approaches are feasible, for example a matched pairs design where each subject reads similar length passages with both sources.

(b) With a test we can determine whether the means for the *populations* are likely to differ, not just the samples. A test also gives us the strength of evidence for the result.

(c) A confidence interval for the difference in means gives an estimate of how large the effect might be, not just whether there is one. Also, like a test, a confidence interval allows us to generalize to the population.

(d) The first quotations only tells us whether the results are significant and not which method is faster. The second quotation gives additional information to show the mean reading time was fastest for the print book and slowest for the Kindle, with the iPad in between.

4.157 (a) If we let p_F and p_M represent the proportion of rats showing compassion between females and males, respectively, we have

$$H_0: \quad p_F = p_M$$
$$H_a: \quad p_F \neq p_M$$

(b) The null hypothesis value for the difference in proportions is $p_F - p_M = 0$. Since 0 is not in the confidence interval, we reject H_0 and find, at the 5% level, that there is evidence of a difference in proportions of compassionate rats between the two genders.

(c) Since we are estimating $p_F - p_M$ and all plausible values are positive, we have strong evidence that $p_F - p_M > 0$ which means $p_F > p_M$. This indicates that female rats are more likely to show compassion.

4.159 (a) This is an observational study. Randomization was used to select the sample of 50 stocks, so there should be no bias.

(b) Number the stocks 001 to 500, put the numbers in a hat, draw out 50 to identify the sample, or use a random number generator.

(c) Most changes are near zero (most slightly above), with a few outliers in both directions.

(d) We see that the mean is $\bar{x} = 0.124$ and the standard deviation is $s = 0.99$. The five number summary is $(-3.27, -0.03, 0.11, 0.32, 4.86)$.

(e) Using the 2.5%-tile and 97.5%-tiles from a bootstrap distribution (shown below) of 1000 means for samples of size 50, we obtain the 95% confidence interval (-0.142, 0.403). We are 95% sure that the average stock price change for S&P 500 stocks during this period was between $-\$0.142$ and $\$0.403$.

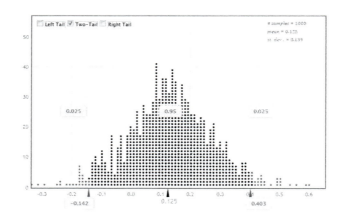

(f) Since the question does not specify a particular direction, the hypotheses are $H_0 : \mu = 0$ vs $H_a : \mu \neq 0$. Since $\mu = 0$ falls within the 95% confidence bounds of part (e) we should not reject H_0 at a 5% level. We do not find evidence that the mean change is different from zero.

(g) To test for a positive mean change the hypotheses are $H_0 : \mu = 0$ vs $H_a : \mu > 0$. In one randomization distribution (shown below) of means for 1000 samples of size 50 when $\mu = 0$, we find 185 values at (or beyond) the observed mean of 0.124. This gives a one-tailed p-value of 0.185, which is not small, so we do not reject H_0. We do not have enough evidence to show that the mean price change for all S&P 500 stocks during this period was positive.

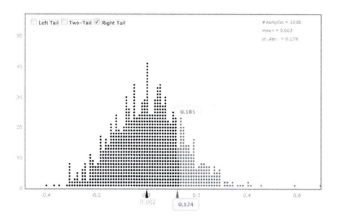

(h) Since the decision was to not reject H_0, only a Type II error is possible. To see if that error was made, we could find the mean price change for all 500 stocks during the period and see if it really was positive.

4.161 (a) We are testing $H_0 : \mu = 200$ vs $H_a : \mu \neq 200$, where μ represents the mean gestation time, in days, of all mammals. The mean gestation in the sample is $\overline{x} = 194.3$ days. We use *StatKey* or other technology to create a randomization distribution such as the one shown below. For this two-tailed test, the proportion of samples beyond $\overline{x} = 194.3$ in this distribution gives a p-value of $2 \cdot 0.388 = 0.776$. This is not a small p-value, so we do not reject H_0. There is not sufficient evidence to say that mean gestation time is different from 200 days.

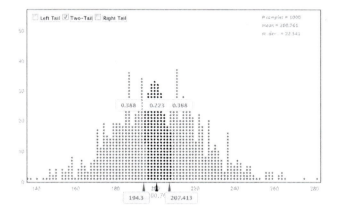

(b) Since we do not reject $\mu = 200$ as a plausible value of the population mean in the hypothesis test in part (a), 200 would be included in a 95% confidence interval for the mean gestation time of mammals.

4.163 (a) We let μ_I and μ_C represent the mean score on the HRSIW subtest for kindergartners who get iPads and kindergartners who don't get iPads, respectively. The hypotheses are:

$$H_0 : \quad \mu_I = \mu_C$$
$$H_a : \quad \mu_I > \mu_C$$

(b) Since the p-value (0.006) is very small, we reject H_0. There is evidence that the mean score on the HRSIW subtest for kindergartners with iPads is higher than the mean score for kindergartners without iPads. The results are statistically significant.

(c) The school board member could be arguing whether a 2 point increase in the mean score on one subtest really matters very much (is *practically* significant). Even though this is a statistically significant difference, it might not be important enough to justify the considerable cost of supplying iPad's to all kindergartners.

4.165 (a) If there is no effect due to food choices and $\alpha = 0.01$, a Type I error should occur on about 1% of tests. We see that 1% of 133 is 1.33, so we would expect one or two tests to show significant evidence just by random chance.

(b) When testing more than 100 different foods, making at least one Type I error is fairly likely.

(c) No, even if the proportion of boys is really higher for mothers who eat breakfast cereal, the data were obtained from an observational study and not an experiment. A headline that implies eating breakfast cereal will *cause* an increase in the chance of conceiving a boy is not appropriate without doing an experiment where cereal habits are controlled.

4.167 (a) We should definitely be less confident. If the authors conducted 42 tests, it is likely that some of them will show significance just by random chance even if massage does not have any significant effects. It is possible that the result reported earlier just happened to one of the random significant ones.

(b) Since none of the tests were significant, it seems unlikely that massage affects muscle metabolites.

(c) Now that we know that only eight tests were testing inflammation, and that four of those gave significant results, we can be quite confident that massage does reduce muscle inflammation after exercise. It would be very surprising to see four p-values (out of eight) less than 5% if there really were no effects at all.

Unit B: Essential Synthesis Solutions

B.1 (a) We use a confidence interval since we are estimating the proportion of voters who support the proposal and there is no claimed parameter value to test.

(b) We use a hypothesis test since we are testing the claim $H_0 : p_h = p_a$ vs $H_a : p_h > p_a$.

(c) We use a hypothesis test to test the claim $H_0 : p = 0.5$ vs $H_a : p > 0.5$.

(d) Inference is not relevant in this case, since we have information on the entire populations.

B.3 (a) The p-value is small so we reject a null hypothesis that says the mean recovery time is the same with or without taking Vitamin C.

(b) There are many possible answers. Here's one example of an inappropriate data collection method: Give Vitamin C only to those who have had cold symptoms for a long time. They may have shorter recovery times since the cold is almost over when they start treatment, while those not getting Vitamin C might be at the early stages of their colds.

(c) We need to randomize the assignment of subjects (students with colds) to the two groups. For example, we could flip a coin to determine who gets Vitamin C (heads) and who gets a placebo (tails). Neither the subjects nor the person determining when they are recovered should know which group they are in.

(d) The small p-value indicates we should reject $H_0 : \mu_c = \mu_{nc}$ in favor of the alternative $H_0 : \mu_c < \mu_{nc}$, where μ denotes the mean recovery time from a cold. Thus we have strong evidence that large doses of Vitamin C help reduce the mean time students need to recover from a cold.

B.5 (a) The hypotheses are $H_0 : \mu_{dc} = \mu_w$ vs $H_a : \mu_{dc} > \mu_w$, where μ_{dc} and μ_w are the mean calcium loss after drinking diet cola and water, respectively. The difference in means for the sample is $\bar{x}_{dc} - \bar{x}_w = 56.0 - 49.125 = 6.875$. We use *StatKey* or other technology to construct a randomization distribution, such as the one shown below, for this difference in means test. We find the p-value in this upper-tail test by finding the proportion of the distribution above the sample difference in means of 6.875. For the randomization distribution below this gives a p-value of 0.005 and strong evidence to reject the null hypothesis. We conclude that mean calcium loss for women is higher when drinking diet cola than when drinking water.

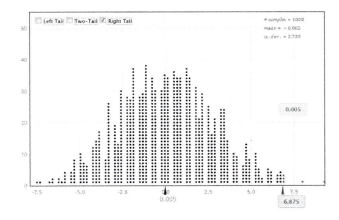

(b) Since we found a significant difference in part (a), we find a confidence interval for the difference in means $\mu_{dc} - \mu_w$ using a bootstrap distribution and either percentiles or the $\pm 2 \cdot SE$ method. For the bootstrap distribution shown below, we see that a 95% confidence interval using percentiles goes from 2.88 to 10.75. We are 95% sure that women who drink 24 ounces of diet cola will increase calcium excretion, on average, between 2.875 and 10.75 milligrams when compared to drinking water. Using $\pm 2 \cdot SE$, the 95% confidence interval is $6.875 \pm 2 \cdot 2.014 = 6.875 \pm 4.028 = (2.85, 10.90)$.

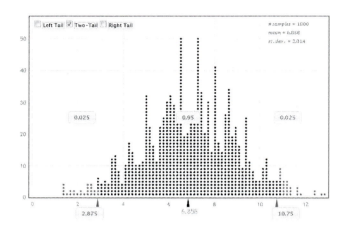

B.7 (a) Roommates are assigned at random, so whether a student has a roommate with a videogame or not is determined at random.

(b) The hypotheses are $H_0 : \mu_v = \mu_n$ vs $H_a : \mu_v < \mu_n$, where μ_v and μ_n are the mean GPA for students whose roommates do and do not bring a videogame, respectively.

(c) At a 5% level, we reject H_0. The mean GPA is lower when a roommate brings a videogame.

(d) Negative differences indicate $\mu_v - \mu_n < 0$ which means $\mu_v < \mu_n$, a lower mean GPA when a roommate brings a videogame. We are 90% sure that students with a roommate who brings a videogame have a mean GPA between 0.315 and 0.015 less than the mean GPA of students whose roommates don't bring a videogame.

(e) At a 5% significance level, we do not reject H_0 when the p-value=0.068. There is not (quite) enough evidence to show that mean GPA among students who don't bring a video game is lower when their roommate does bring one.

(f) At a 5% significance level we reject H_0 when the p-value=0.026. There is enough evidence to show that mean GPA among students who bring a video game is lower when their roommate also brings one.

(g) The effect (reducing mean GPA when a roommate brings a videogame) is larger for students who bring a videogame themselves. Perhaps this makes sense since students who bring a videogame are already predisposed to get distracted by them.

(h) For students who bring a video game to college, we are 90% sure that their mean GPA is lower by somewhere between 0.526 and 0.044 points if their roommate also brings a videogame than if their roommate does not bring a videogame. This interval is similar to the one found in part (d) but is farther in the negative direction.

(i) Having more videogames in the room tends to be associated with lower mean GPA.

(j) There are many possible answers. One possible additional test is to ignore the roommate completely and see if mean GPA in the first semester is lower for students who bring a videogame to college than for students who do not bring one.

B.9 (a) We expect married couples to tend to have similar ages, so we expect a positive correlation between husband and wife ages.

(b) A scatterplot of *Husband* vs *Wife* ages is shown below. We see a strong positive, linear association. The correlation for this sample of data is $r = 0.914$.

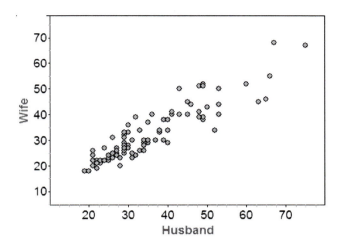

(c) To find a confidence interval for this correlation, we sample (with replacement) from the original data and compute the correlation between husband and wife ages for each bootstrap sample of 105 couples. We repeat this process to generate 5000 values in a bootstrap distribution such as the one shown below. From the percentiles of this distribution of bootstrap correlations, we find a 95% confidence interval to be from 0.877 to 0.945. We are 95% sure that the correlation between husband and wife ages for all recent marriages in this jurisdiction is between 0.877 and 0.945.

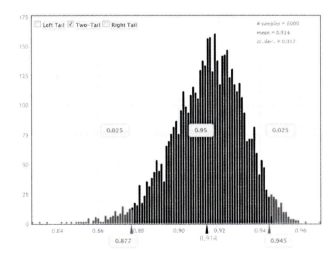

(d) Although we have evidence of a strong, positive correlation between the ages of husbands and wives, the correlation contains no information to help with the previous exercise of deciding whether husbands or wives tends to be older.

Unit B: Review Exercise Solutions

B.11 (a) The relevant population is all American adults, and the parameter we are estimating is p, the proportion of all American adults who believe that violent movies lead to more violence in society. The best point estimate is $\hat{p} = 0.57$.

(b) A 95% confidence interval is

$$
\begin{array}{ccc}
\text{Point Estimate} & \pm & \text{Margin of Error} \\
0.57 & \pm & 0.03 \\
0.54 & \text{to} & 0.60
\end{array}
$$

We are 95% confident that the proportion of all American adults to believe that violent movies lead to more violence in society is between 0.54 and 0.60.

B.13 (a) This is a population proportion so the correct notation is p. Using the data in **Hollywood-Movies2011**, we have $p = 27/136 = 0.1985$.

(b) We expect it to be symmetric and bell-shaped and centered at the population proportion of 0.1985.

B.15 The p-value 0.0004 goes with the experiment showing significantly lower performance on material presented while the phone was ringing. The p-value 0.93 goes with the experiment measuring the impact of proximity of the student to the ringing phone. The p-value of 0.0004 shows very strong evidence that a ringing cell phone in class affects student learning.

B.17 We create a randomization distribution such as the one shown below by randomly assigning the weight gains to the LL and LD groups (9 each) and computing the difference in means $\overline{x}_{LL} - \overline{x}_{LD}$ for each randomization. The value of this statistic for the original sample is $\overline{x}_{LL} - \overline{x}_{LD} = 10.89 - 5.89 = 5.0$. Since none of the 1000 simulated samples gave a difference as extreme as the observed difference of 5.0, we see that the p-value is approximately zero. This provides very strong evidence that mice with a light on at night have a higher mean weight gain after four weeks. (Why this is so is not yet fully understood.)

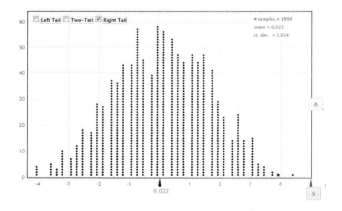

B.19 The point estimate for the population mean is $\overline{x} = 3.04$, so we have

$$
\begin{array}{rcl}
\text{Interval estimate} & = & \text{point estimate} \pm \text{margin of error} \\
& = & \overline{x} \pm \text{margin of error} \\
& = & 3.04 \pm 0.86.
\end{array}
$$

Since $3.04 - 0.86 = 2.18$ and $3.04 + 0.86 = 3.90$, the interval estimate is 2.18 to 3.90. We have some confidence that the mean tip amount for all her deliveries lies somewhere between \$2.18 to \$3.90. For these estimates to be accurate, we need to assume that there is nothing special about the nights included in the sample so that it is representative of all her deliveries. For example, the samples were all from Friday and Saturday nights, so we might not expect the interval to hold for the mean tip on Tuesday nights.

B.21 The best point estimate is $\hat{p} = 0.56$ and a 95% confidence interval is

Point Estimate	$\pm$	Margin of Error
0.56	$\pm$	0.03
0.53	to	0.59

We are 95% confident that the proportion of all American adults who rarely or never go out to the movies is between 0.53 and 0.59. This entire interval is above 50%, so we can be relatively sure that the percentage is greater than 50%.

B.23 (a) Since we are looking at whether smoking has a negative effect, this is a one-tailed test.

(b) The null hypothesis will be that the two proportions are identical $H_0 : p_s = p_{ns}$, and the alternative is that the proportion of successful pregnancies will be less in the smoking group $H_a : p_s < p_{ns}$.

(c) We want the number assigned to each group to match the numbers in the original sample, so 135 women to the smoking group and 543 to the non-smoking. (Both of these values can be found in the two-way table.)

(d) In the original sample, there were 38 successful pregnancies in the 135 women in the smoking group. From the randomization distribution, it appears that about 40 of the 1000 values fall less than or equal to the count of 38 from the original sample, so the best estimate for the p-value is about $40/1000 = 0.04$.

B.25 (a) Sample A. The sample mean in A is around 43, while the sample mean in B is around 47. Sample sizes and variability are similar for both samples.

(b) Sample B. Both samples appear to have a mean near 46, but the variability is smaller in sample B so we can be more sure the mean is below 50. Also, sample B has few values above 50, while sample A has at least 25% of its values above 50 (since $Q_3 > 50$).

(c) Sample A. Both samples appear to have about the same mean and median (near 45) and similar variability, but sample A is based on a much larger sample size, so it would be more unusual to see that many values below 50 if $H_0 : \mu = 50$ were true.

B.27 (a) This is a population proportion so the correct notation is p. We have $p = 32/136 = 0.235$.

(b) Using technology we produce a sampling distribution (shown below) of 1000 sample proportions when samples of size $n = 30$ are drawn from a population with $p = 0.235$. We see that the distribution is relatively symmetric, bell-shaped and centered at the population proportion of 0.235, as we expect. We also see in the figure that the estimated standard error based on these 1000 simulated proportions is 0.08.

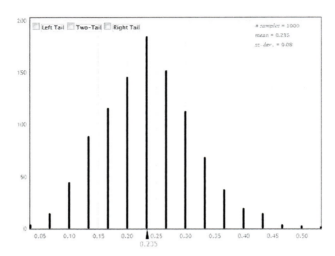

B.29   (a) Approximately the same. We expect both distributions to be approximately symmetric and bell-shaped.

   (b) Different. The sampling distribution is centered at the value of the population parameter, while the bootstrap distribution is centered at the value of the sample statistic.

   (c) Approximately the same. The standard error from the bootstrap distribution gives a good approximation to the standard error for the sampling distribution.

   (d) Different. One value in the sampling distribution represents the statistic from a sample taken (without replacement) from the entire population, while one value in the bootstrap distribution represents the statistic from a sample taken with replacement from the original sample. In both cases, however, we compute the same statistic (mean, proportion, or whatever) and use the same sample size.

   (e) Different. In order to create a sampling distribution, we need to know the data values for the entire population! In order to create a bootstrap distribution, we only need to know the values in one sample. This is what makes the bootstrap method so powerful.

B.31 We use *StatKey* or other technology to generate a bootstrap distribution for the difference in means. For a 90% confidence interval, we keep the middle 90% of values in the bootstrap distribution and cut off 5% in each tail. We see in the figure that the 90% confidence interval for the difference in weekly hours spent exercising is $(-0.92, 6.87)$. We are 90% confident that the difference in mean number of hours spent exercising between male and female college students is between -0.92 hours and 6.87 hours. This means the true difference is likely to be anywhere from females exercising, on average, 0.92 hours more than males to males exercising, on average, 6.87 hours more than females. Notice that the 90% interval calculated here is narrower than the 95% confidence interval (-1.75 to 7.75) calculated earlier using the standard error. This makes sense, since a 95% confidence interval needs to be wider to have a better chance of capturing the true difference.

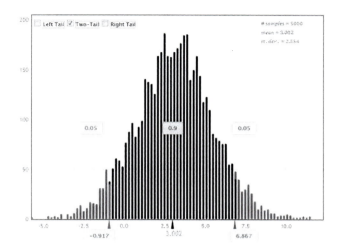

B.33 (a) We put the 10 data values (time immobile) from the sample on 10 slips of paper. We mix them up, select one, record the value, and put it back. Do this 10 times. The statistic recorded is the mean of the 10 values obtained that way.

(b) A 95% confidence interval is $\bar{x} \pm 2 \cdot SE = 135 \pm 2(6) = 135 \pm 12$. We are 95% sure that the mean time immobile for mice who were depressed and then received a shot of ketamine is between 123 and 147 seconds.

(c) The value of 160 is outside of the confidence interval from part (b), so is not a plausible mean value for immobile time of mice treated with ketamine. It appears that, on average, ketamine reduces this measure of depression in mice.

B.35 (a) The mean and standard deviation of internet access rates for all countries are population parameters so the correct notation is μ and σ. Using the (non-missing) data cases in **AllCountries**, we have $\mu = 28.96\%$ and $\sigma = 26.31$.

(b) The countries with the highest internet access rates are Norway and Iceland, with 90.5% of their populations having access to the internet. The lowest rate is Myanmar, at 0.2%. Answers will vary for the percentage in your country, depending on where you live.

(c) A sampling distribution with means for 1000 samples of size $n = 10$ taken from the internet access rates for the population of all countries is shown below. We see that the distribution is symmetric, bell-shaped, and centered at the population mean of 29 percent, as we expect. The standard deviation of these 1000 sample means is 8.1, so we estimate the standard error for mean internet access rates based on samples of 10 countries to be $SE \approx 8.1$.

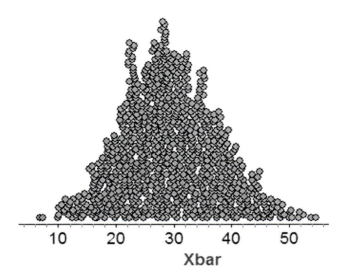

B.37 When the results of the study are not statistically significant, we fail to reject the null hypothesis (in this case that heavy cell phone use is unrelated to developing bring cancer). But that does not mean we "accept" that H_0 *must* be true, we just lack sufficiently convincing evidence to refute it.

(a) By not rejecting H_0 we are saying it is a plausible option, so it might be true that heavy cell phone use has no effect on developing brain cancer.

(b) There is some evidence in the sample that heavy cell phone users have a higher risk of developing brain cancer, just not enough to be considered statistically significant. Failing to reject H_0 means either H_0 or H_a might still be true. Note that any confidence interval that includes zero, contains both positive and negative values as plausible options. Hence the authors tell us that the question "remains open".

(c) We note that this study was an observational study and not an experiment, and thus, even if the results had been statistically significant, we would not make a cause/effect conclusion about this relationship. However, a *lack* of significant results does not rule this out as a plausible option.

B.39 The point estimate from the original sample is $\hat{p} = 0.45$. The figure shows the proportions for 10,000 bootstrap samples of size 147,291 (the original sample size) when the proportion is 0.45. We see that the distribution is relatively symmetric and bell-shaped and we see in the upper-right of the figure that the standard error based on the bootstrap distribution is $SE = 0.001$. (Looking at the distribution, an estimate of 0.0015 or anything in between would also be reasonable.) The 95% confidence interval is

$$
\begin{array}{ccc}
\hat{p} & \pm & 2 \cdot SE \\
0.45 & \pm & 2 \cdot 0.001 \\
0.45 & \pm & 0.002 \\
0.448 & \text{to} & 0.452
\end{array}
$$

For a sample as large as 147,291 the margin of error (0.002 or 0.2%) is quite small. We can be 95% confident that the proportion of all American adults to get health insurance from an employer is between 44.8% and 45.2%.

B.41 This is a hypothesis test for a single proportion and we define p to be the proportion of times the lie detector will report lying when a stressed individual is telling the truth. The hypotheses are:

$$H_0 : \quad p = 0.5$$
$$H_a : \quad p > 0.5$$

We create a randomization distribution of sample proportions (shown below) using $p = 0.5$ and samples of size $n = 48$. The statistic for the original sample is $\hat{p} = 27/48 = 0.563$ and this is a right-tailed test so the p-value is the proportion of samples with proportions beyond 0.563. For the randomizations distribution below this gives a p-value of 0.229 which is not small. We do not reject H_0, and do not find evidence that the software gives inaccurate results more than half the time in this situation .

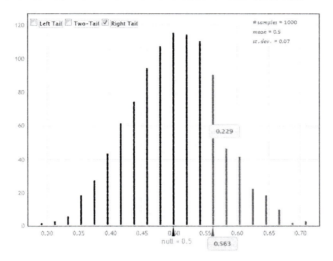

B.43 Using *StatKey* or other technology we construct a bootstrap distribution based on the mean commute distances for 5000 samples of size $n = 500$ taken (with replacement) from the original **CommuteAtlanta** distances. For the distribution shown below, the 5%-tile is 17.15 and the 95%-tile is 19.15 so the 90% confidence interval is (17.15, 19.15). We are 90% sure that the average distance to work for all commuters in metropolitan Atlanta is between 17.15 miles and 19.15 miles.

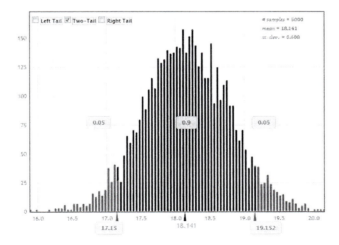

B.45   (a) Let ρ denote the correlation between pH and mercury in all Florida lakes. The question of interest is whether or not this correlation is negative, so we use a one-tailed test, with hypotheses $H_0 : \rho = 0$ vs $H_a : \rho < 0$.

   (b) We want to find a point that has roughly 30% of the randomization distribution in the lower tail below it. This should occur somewhere between $r = -0.20$ and $r = -0.10$, perhaps $r \approx= -0.15$. (It is difficult to determine this point very precisely from the plot.)

   (c) We want to find a point that has only about 1% of the randomization distribution in the lower tail below it. This should occur around $r \approx -0.50$.

B.47   (a) In a Type I error, we conclude that treated wipes prevent infection, when actually they don't.

   (b) In a Type II error, we conclude that treated wipes are not shown to be effective, when actually they help prevent infections.

   (c) A smaller significance level means we need more evidence to reject H_0. We would want a smaller significance level in the second situation (harmful side effects) so it has to be very clear that the treated wipes help prevent infection, since we don't want to put people at risk for side effects if the benefit isn't definite.

   (d) The p-value (0.32) is not small, so we do not reject H_0. The study does not provide sufficient evidence to show that treated wipes are more effective at reducing the proportion of infected babies than sterile wipes.

   (e) Not necessarily. The results of the test are inconclusive when the p-value is not small. Either H_0 or H_a could still be valid, so the treated wipes might help prevent infections and the study just didn't accumulate enough evidence to verify it.

B.49 We use technology to create a randomization distribution (such as shown below) and then check to see how extreme the original sample statistic of $r = -0.575$ is in that distribution. None of the randomization statistics in this distribution are beyond (less than since this is a lower-tail test) the observed $r = -0.575$. Thus the p-value ≈ 0.000. Based on this very small p-value, we have strong evidence to reject H_0 and conclude that there is negative correlation between pH levels and the amount of mercury in fish of Florida lakes. More acidic lakes tend to have higher mercury levels in the fish.

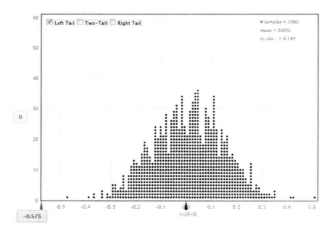

B.51 For one set of 1000 randomization samples (shown below), 242 of the sample correlations are more than the observed $r = 0.279$ which gives a p-value of 0.242. This p-value is not smaller than any reasonable significance level, so we do not have sufficient evidence to reject H_0. Based on this sample of 8 patients, we can not conclude that there must be a positive association between heart rate and systolic blood pressure for 55 year-old patients at this ICU.

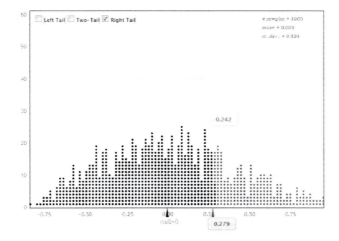

B.53 The null hypothesis for this correlation test is that the correlation is zero. To create the randomization samples, we match the null hypothesis. In this situation, that means height and salary are completely unrelated. We might randomly scramble the height values and assign them to the original subjects/salaries. Compute the correlation, r, between those random heights and the actual salaries.

B.55 The null hypothesis for this difference in means test is that the means of the two groups are the same. To create the randomization samples, we match the null hypothesis. In this situation, that means whether or not a customer is approached has no effect on sales. We might randomly scramble the labels for type of store ("approach" or "not approach") and assign them to the actual sales values. Compute the difference in the mean sales, $\overline{x}_a - \overline{x}_{na}$, between the stores assigned to the "approach" group and those randomly put in the "not approach" group. Other methods, for example sampling with replacement from the pooled sales values to simulate new samples of "approach" and "not approach" sales, are also acceptable.

Section 5.1 Solutions

5.1 The area for values below 25 is clearly more than half the total area, but not as much as 95%, so 62% is the best estimate.

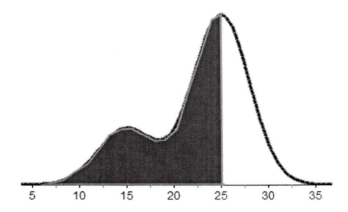

5.3 Almost all of the area under the density curve is between 10 and 30, so 95% is the best estimate.

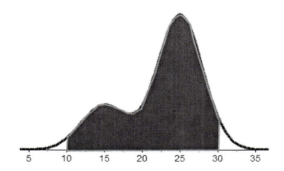

5.5 The plots below show the three required regions as areas in a $N(0,1)$ distribution. We see that the areas are

   (a) 0.8508

   (b) 0.9332

   (c) 0.1359

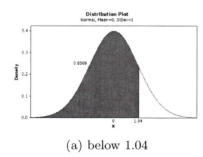   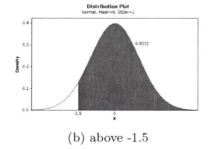   

      (a) below 1.04   (b) above -1.5   (c) between 1 and 2

5.7 The plots below show the three required regions as areas in a N(0,1) distribution. We see that the areas are

(a) 0.982

(b) 0.309

(c) 0.625

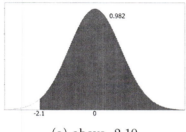
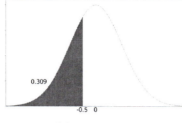
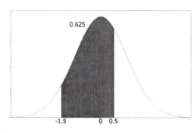

   (a) above -2.10   (b) below -0.5   (c) between -1.5 and 0.5

5.9 The plots below show the required endpoint(s) for a $N(0,1)$ distribution. We see that the endpoint z is

(a) -1.282

(b) -0.8416

(c) ± 1.960

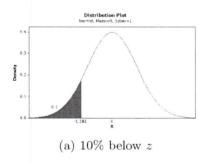
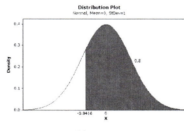
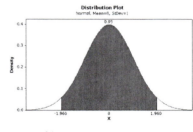

  (a) 10% below z 80% above z 95% between $-z$ and $+z$

5.11 The plots below show the required endpoint(s) for a N(0,1) distribution. We see that the endpoint z is

(a) -1.28

(b) 0.385

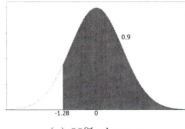
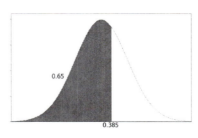

   (a) 90% above z (b) 65% below z

5.13 The plots below show the three required regions as areas on the appropriate normal curve. Using technology, we can find the areas directly, and we see that the areas are as follows. (If you are using a paper table, you will need to first convert the values to the standard normal using z-scores.) For additional help, see the online supplements.

(a) 0.691

(b) 0.202

(c) 0.643

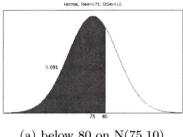
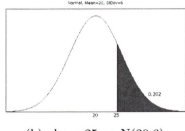
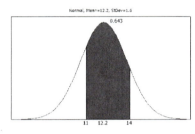

(a) below 80 on N(75,10)   (b) above 25 on N(20,6)   (c) between 11 and 14 on N(12.2, 1.6)

5.15 The plots below show the three required regions as areas on the appropriate normal curve. Using technology, we can find the areas directly, and we see that the areas are as follows. (If you are using a paper table, you will need to first convert the values to the standard normal using z-scores.) For additional help, see the online supplements.

(a) 0.023

(b) 0.006

(c) 0.700

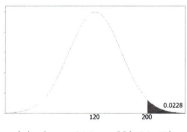
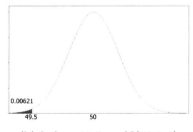
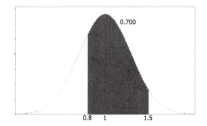

(a) above 200 on N(120,40)   (b) below 49.5 on N(50,0.2)   (c) between 0.8 and 1.5 on N(1,0.3)

5.17 The plots below show the required endpoint(s) for the given normal distribution. Using technology, we can find the endpoints directly, and we see that the requested endpoints are as follows. (If you are using a paper table, you will need to find the endpoints on a standard normal table and then convert them back to the requested normal.) For additional help, see the online supplements.

(a) 59.3

(b) 2.03

(c) 60.8 and 139.2. Notice that this is very close to our rough rule that about 95% of a normal distribution is within 2 standard deviations of the mean.

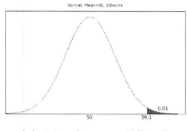
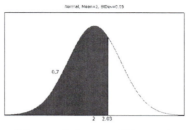
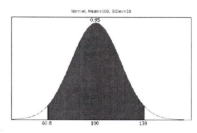

(a) 0.01 above on N(50,4)   (b) 0.70 below on N(2,0.05)   (c) 95% between on N(100,20)

5.19 The plots below show the required endpoint(s) for the given normal distribution. Using technology, we can find the endpoints directly, and we see that the requested endpoints are as follows. (If you are using a paper table, you will need to find the endpoints on a standard normal table and then convert them back to the requested normal.) For additional help, see the online supplements.

(a) 110

(b) 9.88

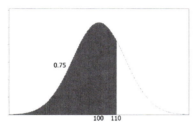
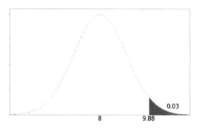

(a) 0.75 below on N(100,15)   (b) 0.03 above on N(8,1)

5.21 We standardize the endpoint of 40 using a mean of 48 and standard deviation of 5 to get

$$z = \frac{x - \mu}{\sigma} = \frac{40 - 48}{5} = -1.6$$

The graphs below show the lower tail region on each normal density. The shaded area in both curves is 0.0548.

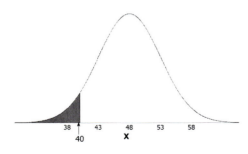
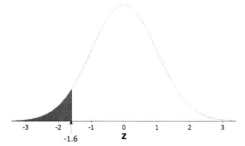

5.23 We use technology to find the endpoint on a standard normal curve that has 5% above it. The graph below shows this point is $z = 1.64$. We then convert this to a N(10,2) endpoint with

$$x = \mu + z \cdot \sigma = 10 + 1.64 \cdot 2 = 13.3$$

Note: We could also use technology to find the N(10,2) endpoint directly. The graphs below show the upper 5% region on each normal density.

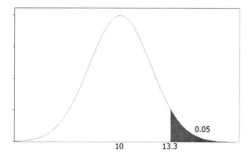
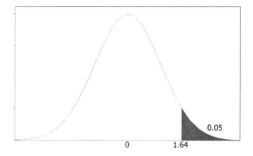

5.25 Using technology to find the endpoint for a standard normal density, we see that it is $z = -1.28$ (as shown in the figure). We convert this to N(500,80) using

$$x = \mu + z \cdot \sigma = 500 - 1.28 \cdot 80 = 397.6$$

Note: We could also use technology to find the N(500,80) endpoint directly. The graphs show the lower 10% region on each normal density.

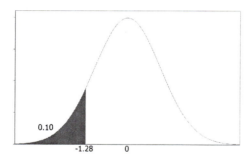
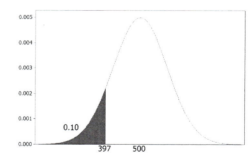

5.27 We convert the standard normal endpoints to N(100,15) using

$$x = 100 + 1 \cdot 15 = 115 \qquad \text{and} \qquad x = 100 + 2 \cdot 15 = 130$$

The graphs below show the region on each normal density. We see that the area is identical in both and is 0.1359.

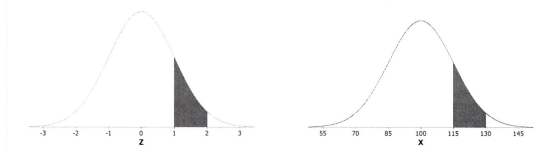

5.29 A $N(0.275, 0.02)$ curve is centered at its mean, 0.275. The values two standard deviations away, 0.235 and 0.315, are labeled so that approximately 95% of the area falls between these two values. They should be out in the tails, with only about 2.5% of the distribution beyond them on each side. See the figure.

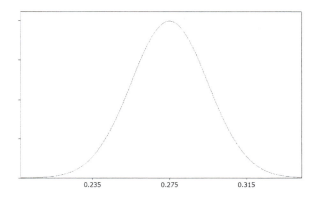

5.31 Using technology we can find the area above 0.30 on a $N(0.275, 0.02)$ curve directly. We see in the figure that the area is 0.106.

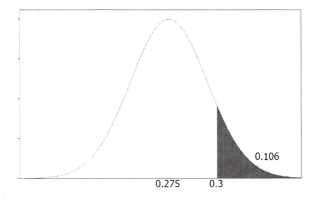

Alternately, we can compute a z-score

$$z = \frac{0.30 - 0.275}{0.02} = 1.25$$

Using technology or a table the area above 1.25 on a standard normal curve is 0.106, the same as the area we found above. Thus about 10.6% of samples of 500 US adults will contain more than 30% with at least a bachelor's degree.

5.33 The plots below show the required endpoint(s) and/or probabilities for the given normal distributions. Note that a percentile always means the area to the left. Using technology, we can find the endpoints and areas directly, and we obtain the answers below. (Alternately, we could convert to a standard normal and use the standard normal to find the equivalent area.)

(a) The area below 450 is 0.353, so that point is the 35th percentile of a N(492,111) distribution.

(b) The point where 90% of the scores are below it is a score of 634.

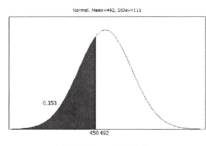

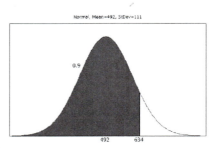

(a) 450 on Writing   (b) 90th percentile Writing

5.35   (a) Using technology we find the area between 68 inches and 72 inches in a $N(70,3)$ distribution. We see in the figure that the area is 0.495.

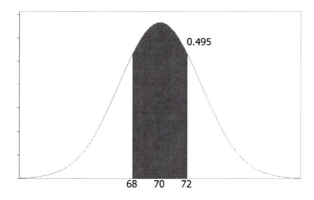

Alternately, we can compute z-scores:

$$z = \frac{68-70}{3} = -0.667 \qquad z = \frac{72-70}{3} = 0.667$$

Using technology or a table the area between -0.667 and 0.667 on a standard normal curve is 0.495, matching what we see directly. About 49.5% (or almost exactly half) of US men are between 68 and 72 inches tall.

(b) Using technology we find an endpoint for a $N(70, 3)$ distribution that has an area of 0.10 below it. We see in the figure that the height is about 66.2 inches.

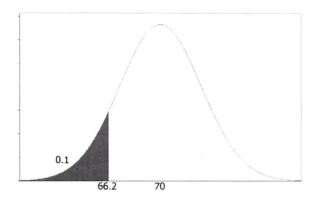

Alternately, we can use technology or a table to find the 10%-tile for a standard normal distribution, $z = -1.28$ and convert this value to the $N(70, 3)$ scale, $Height = 70 - 1.28(3) = 66.2$. Again, of course, we arrive at the same answer using either approach. A US man whose height puts him at the 10th percentile is 66.2 inches tall or about 5'6".

5.37 We use technology to find the points on a N(21.97, 0.65) curve that have 25% and 75%, respectively, of the distribution below them. These points are at $Q_1 = 21.53$ minutes and $Q_3 = 22.41$ minutes.

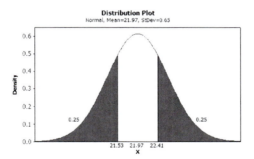

Alternately, we can use technology or a table to find the 25% and 75% endpoints for a standard normal distribution, $z = \pm 0.6745$. We then convert these endpoints to corresponding points on a $N(21.97, 0.65)$ distribution.

$$Q_1 = 21.97 - 0.6745 \cdot 0.65 = 21.53$$
$$Q_3 = 21.97 + 0.6745 \cdot 0.65 = 22.41$$

5.39 We use technology to determine the answers. We see in the figure that the results are:

(a) 0.0509 or 5.09% of students scored above a 90.

(b) 0.138 or 13.8% of students scored below a 60.

(c) Students with grades below 53.9 will be required to attend the extra sessions.

(d) Students with grades above 86.1 will receive a grade of A.

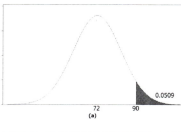

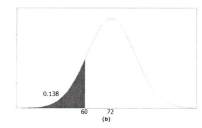

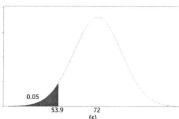

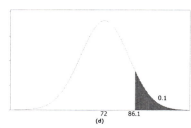

5.41 (a) For any $N(\mu, \sigma)$ distribution, when we standardize $\mu - 2\sigma$ and $\mu + 2\sigma$ we must get $z = -2$ and $z = +2$. For example, if we use $N(100, 20)$ the interval within two standard deviations of the mean goes from 60 to 140.

$$z = \frac{60 - 100}{20} = -2 \qquad \text{and} \qquad z = \frac{140 - 100}{20} = +2$$

Using technology, the area between -2 and $+2$ on a standard normal curve is 0.954.

(b) Similar to part (a), if we go just one standard deviation in either direction the standardized z-scores will be $z = -1$ and $z = +1$. Using technology, the area between -1 and $+1$ on a standard normal curve is 0.683.

(c) Similar to part (a), if we go three standard deviations in either direction the standardized z-scores will be $z = -3$ and $z = +3$. Using technology, the area between -3 and $+3$ on a standard normal curve is 0.997.

(d) The percentages within one, two, or three standard deviations of the mean, roughly 68% between $\mu \pm \sigma$, roughly 95% between $\mu \pm 2\sigma$, and roughly 99.7% between $\mu \pm 3\sigma$, should hold for any normal distribution since the standardized z-scores will always be $z = \pm 1$ or $z = \pm 2$ or $z = \pm 3$, respectively.

Section 5.2 Solutions

5.43 (a) For 86% confidence we need $(1 - 0.86)/2 = 0.07$ in each tail. Using technology the standard normal points with this area are $\pm z^* = \pm 1.476$.

(b) For 94% confidence we need $(1 - 0.94)/2 = 0.03$ in each tail. Using technology the standard normal points with this area are $\pm z^* = \pm 1.881$.

(c) For 96% confidence we need $(1 - 0.96)/2 = 0.02$ in each tail. Using technology the standard normal points with this area are $\pm z^* = \pm 2.054$.

5.45 For a 95% confidence interval, we have $z^* = 1.96$. The confidence interval is:

$$
\begin{array}{rcl}
\text{Sample statistic} & \pm & z^* \cdot SE \\
\overline{x} & \pm & z^* \cdot SE \\
72 & \pm & 1.96(1.70) \\
72 & \pm & 3.332 \\
68.668 & \text{to} & 75.332.
\end{array}
$$

5.47 For a 99% confidence interval, we have $z^* = 2.576$. The confidence interval is:

$$
\begin{array}{rcl}
\text{Sample statistic} & \pm & z^* \cdot SE \\
\hat{p} & \pm & z^* \cdot SE \\
0.78 & \pm & 2.576(0.03) \\
0.78 & \pm & 0.077 \\
0.703 & \text{to} & 0.857.
\end{array}
$$

5.49 For a 95% confidence interval, we have $z^* = 1.96$. The confidence interval is:

$$
\begin{array}{rcl}
\text{Sample statistic} & \pm & z^* \cdot SE \\
(\overline{x}_1 - \overline{x}_2) & \pm & z^* \cdot SE \\
(256 - 242) & \pm & 1.96(6.70) \\
14 & \pm & 13.132 \\
0.868 & \text{to} & 27.132.
\end{array}
$$

5.51 For this test of a proportion we compare the sample proportion to the hypothesized proportion and divide by the standard error.

$$
z = \frac{\text{Sample proportion} - \text{Null proportion}}{SE} = \frac{0.235 - 0.25}{0.018} = -0.83
$$

5.53 For this test of a mean we compare the sample mean to the hypothesized mean and divide by the standard error.

$$
z = \frac{\text{Sample mean} - \text{Null mean}}{SE} = \frac{11.3 - 10}{0.10} = 13.0
$$

5.55 The relevant statistic here is the difference in means. From the null hypothesis, we see that $\mu_1 - \mu_2 = 0$. We have:

$$
z = \frac{\text{Sample difference in means} - \text{Null difference in means}}{SE} = \frac{(35.4 - 33.1) - 0}{0.25} = 9.2.
$$

5.57 (a) Using technology or a table the standard normal area below $z = -1.08$ is 0.140. The p-value for a lower tail test is 0.140.

(b) Using technology or a table the standard normal area above $z = 4.12$ is 0.0000189. We don't really need technology or a table to find this p-value. The standardized test statistic is so large (more than four standard deviations above the mean) that the area beyond it is essentially zero. The p-value for this test is approximately zero.

(c) Using technology or a table the standard normal area below $z = -1.58$ is 0.0571. Double this to find the p-value for a two-tail test, $2 \cdot 0.0571 = 0.1142$.

These three p-values are shown as areas below.

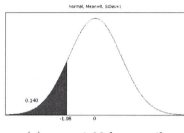

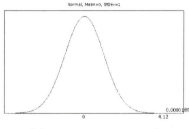

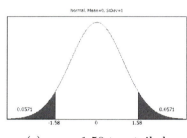

(a) $z = -1.08$ lower tail   (b) $z = 4.12$ upper tail   (c) $z = -1.58$ two-tailed

5.59 To find a confidence interval using the normal distribution, we use

$$\text{Sample statistic} \pm z^* \cdot SE.$$

We are finding a confidence interval for a proportion, so the statistic from the original sample is $\hat{p} = 0.60$. For a 99% confidence interval, we use $z^* = 2.576$ and we have $SE = 0.015$. Putting this information together, we have

$$
\begin{array}{rcl}
\hat{p} & \pm & z^* \cdot SE \\
0.60 & \pm & 2.576 \cdot (0.015) \\
0.60 & \pm & 0.039 \\
0.561 & \text{to} & 0.639
\end{array}
$$

We are 99% sure that the percent of all air travelers who prefer a window seat is between 56.1% and 63.9%.

5.61 (a) To find a confidence interval using the normal distribution, we use

$$\text{Sample statistic} \pm z^* \cdot SE.$$

We are finding a confidence interval for a mean, so the relevant statistic from the original sample is $\overline{x} = 5.2$. For a 95% confidence interval, we use $z^* = 1.960$ and we have $SE = 0.7$. Putting this information together, we have

$$
\begin{array}{rcl}
\overline{x} & \pm & z^* \cdot SE \\
5.2 & \pm & 1.960 \cdot (0.7) \\
5.2 & \pm & 1.37 \\
3.83 & \text{to} & 6.57
\end{array}
$$

We are 95% sure that, before the smoke-free legislation, childhood hospital admissions for asthma were increasing at an average rate of between 3.8% per year and 6.6% per year.

(b) To find a confidence interval using the normal distribution, we use

$$\text{Sample statistic } \pm z^* \cdot SE.$$

We are finding a confidence interval for a mean, so the relevant statistic from the original sample is $\bar{x} = -18.2$. Notice that the change is negative since admissions are decreasing. For a 95% confidence interval, we use $z^* = 1.960$ and we have $SE = 1.79$. Putting this information together, we have

$$
\begin{array}{rcl}
\bar{x} & \pm & z^* \cdot SE \\
-18.2 & \pm & 1.960 \cdot (1.79) \\
-18.2 & \pm & 3.51 \\
-21.71 & \text{to} & -14.69
\end{array}
$$

We are 95% sure that, after the smoke-free legislation, childhood hospital admissions for asthma were *decreasing* at an average rate of between 14.7% per year and 21.7% per year.

(c) This is an observational study. Note that no one randomly assigned some months to have the law and others to not have the law.

(d) Although the results are very compelling, we cannot conclude that the legislation is causing the change in asthma admissions. Because this was not an experiment, there might be confounding variables that happened at the same time as the smoke-free legislation was passed.

5.63 To find a confidence interval using the normal distribution, we use

$$\text{Sample statistic } \pm z^* \cdot SE.$$

We are finding a confidence interval for a difference in proportions, so the relevant statistic from the original sample is $\hat{p}_Q - \hat{p}_R = 0.42 - 0.15 = 0.27$. For a 99% confidence interval, we use $z^* = 2.576$ and we have $SE = 0.07$. Putting this information together, we have

$$
\begin{array}{rcl}
(\hat{p}_Q - \hat{p}_R) & \pm & z^* \cdot SE \\
(0.42 - 0.15) & \pm & 2.576 \cdot (0.07) \\
0.27 & \pm & 0.18 \\
0.09 & \text{to} & 0.45
\end{array}
$$

We are 99% sure that the proportion of words remembered correctly will be between 0.09 and 0.45 higher for people who incorporate quizzes into their study habits.

5.65 If we let p represent the proportion of all American adults who have gone a week without using cash to pay for anything, the hypotheses are:

$$
\begin{array}{rl}
H_0 : & p = 0.40 \\
H_a : & p > 0.40
\end{array}
$$

The relevant sample statistic is $\hat{p} = 0.43$ and the null parameter is 0.40. The standard error is $SE = 0.016$. We use this information to compute the test statistic:

$$z = \frac{\text{Sample statistic} - \text{Null parameter}}{SE} = \frac{0.43 - 0.40}{0.016} = 1.875.$$

This is an upper tail test so the p-value is the area in the standard normal that is above 1.875. We see that the p-value is 0.0304. At a 5% significance level, we find evidence that the proportion not using cash is greater than 40%.

5.67 Under the null hypothesis the randomization distribution should be centered at zero with an estimated standard error of 0.393. The standardized test statistic is

$$z = \frac{0.79 - 0}{0.393} = 2.01$$

Since this is an upper tail test, the p-value is the area in a standard normal distribution above 2.01. Technology or a table shows this area to be 0.022 which is a fairly small p-value (significant at a 5% level). This gives support for the alternative hypothesis that students who smile during the hearing will, on average, tend to get more leniency than students with a neutral expression.

5.69 (a) Using technology to fit a least squares line, the prediction equation is $\widehat{Price} = 30.495 - 0.219 \cdot Miles$. The slope for this sample is -0.219, meaning the price drops by roughly \$219 for every extra 1,000 miles a Mustang has been driven.

   (b) To find a confidence interval for the slope we use technology to select bootstrap samples with replacement from the original 25 Mustangs, keeping track of the slope for each sample. A dotplot of slopes for one set of 5000 bootstrap samples is shown below.

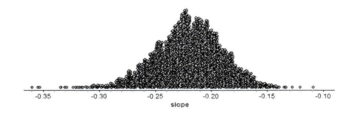

Based on the standard deviation of these 5000 sample slopes, we estimate the standard error of the slope for this regression model to be $SE = 0.0295$. The bootstrap distribution of slopes looks reasonably normal so, for 98% confidence, we use the 1%-tile and 99%-tile from a standard normal density, $z^* = \pm 2.326$. To compute the confidence interval

$$-0.219 \pm 2.326 \cdot 0.0295 = -0.219 \pm 0.069 = (-0.288, -.150)$$

We are 98% sure that the slope of the regression line to predict *Price* based on *Miles* for Mustangs is between -0.288 and -0.150. While our sample slope indicates an average decrease of about \$219 for every extra 1,000 miles of driving, it is quite plausible that this decrease could be anywhere from \$288 to \$150. Notice that the confidence interval (and in fact the entire bootstrap distribution) contains only negative values, so we can be very confident that prices go down as mileage increases.

5.71 (a) For each bootstrap, sample 1,917 values with replacement from a file that has 29% or 556 "yes" values and 71% or 1,361 "no" values (or any set that has 29% "yes" values). Find the proportion of "yes" responses in each sample. Repeat many times and compute the standard deviation of the sample proportions.

   (b) The bootstrap distribution of sample proportions appears to have a normal distribution. For a 99% confidence interval, the standard normal endpoint (leaving 0.005 area in the upper tail) is $z^* = 2.576$. Thus the 99% confidence interval is

$$0.29 \pm 2.576 \cdot 0.0102 = 0.29 \pm 0.026 = (0.264, 0.316)$$

We are 99% sure that between 26.4% and 31.6% of cell phone users (in 2010) have downloaded at least one app to their phone.

5.73 (a) The randomization distribution is formed under the assumption that the null hypothesis, $H_0 : \rho = 0$, is true. A standard error computed from this distribution might not be a good estimate of the standard error for correlations when the "true" correlation is around 0.521.

(b) Instead of using a randomization distribution to estimate the standard error, the student should find correlations for bootstrap samples (with replacement) from the original sample, then estimate the standard error based on the standard deviation of those bootstrap correlations.

(c) One set of 1000 bootstrap correlations is shown in the dotplot below.

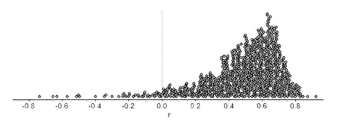

Based on the standard deviation of these bootstrap correlations, we estimate the standard error for the sample correlations to be about $SE = 0.235$ (which is quite close to the standard deviation of 0.22 from the randomization distribution). For a 90% confidence interval, the standard normal endpoint is $z^* = 1.645$. This gives a confidence interval for the correlation of

$$0.521 \pm 1.645 \cdot 0.235 = 0.521 \pm 0.387 = (0.134, 0.908)$$

Thus we are 90% sure that the correlation between the malevolence rating of uniforms and the number of penalty minutes for NHL teams is somewhere between 0.134 and 0.908. Note that this is a very wide interval, although it does contain only positive values for the correlation.

(d) The bootstrap distribution of correlations shows a clear left skew. A normal distribution is probably not appropriate in this case. We should question the validity of the confidence interval found in (c).

5.75 (a) For a 99% confidence interval the standard normal value leaving 0.5% in each tail is $z^* = 2.576$. From the bootstrap distribution we estimate the standard error of the correlations to be 0.205. Thus the 99% confidence interval based on a normal distribution would be

$$0.37 \pm 2.576 \cdot 0.205 = 0.37 \pm 0.528 = (-0.158, 0.898)$$

(b) The bootstrap distribution of correlations is somewhat right skewed, while the normal-based interval assumes the distribution is symmetric and bell-shaped.

Section 6.1 Solutions

6.1   (a) The sample proportions will have a mean of 0.25 and a standard error of

$$SE = \sqrt{\frac{p(1-p)}{n}} = \sqrt{\frac{0.25(1-0.25)}{50}} = 0.061$$

  (b) Since $np = 50(0.25) = 12.5$ and $n(1-p) = 50(0.75) = 37.5$ are both larger than 10, the normal distribution applies and the distribution of sample proportions will be N(0.25, 0.061). We use technology to sketch the graph, or we can sketch it by hand, noting that the sample proportions will be centered at the population proportion of 0.25 and roughly 95% of the distribution lies within two standard deviations on either side of the center. This goes between $0.25 \pm 2(0.061)$ or 0.128 to 0.372.

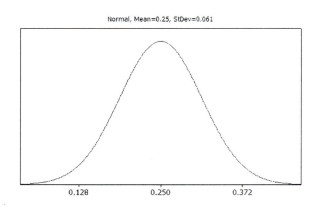

Normal, Mean=0.25, StDev=0.061

           0.128   0.250   0.372

6.3   (a) The sample proportions will have a mean of 0.90 and a standard error of

$$SE = \sqrt{\frac{p(1-p)}{n}} = \sqrt{\frac{0.90(1-0.90)}{60}} = 0.039$$

  (b) We check: $np = 60(0.90) = 54$ and $n(1-p) = 60(0.10) = 6$. Since one of these is less than 10, the sample size is not large enough for a normal distribution to provide a good approximation.

6.5   (a) The sample proportions will have a mean of 0.08 and a standard error of

$$SE = \sqrt{\frac{p(1-p)}{n}} = \sqrt{\frac{.08(0.92)}{300}} = 0.016$$

  (b) Since $np = 300(0.08) = 24$ and $n(1-p) = 300(0.92) = 276$ are both larger than 10, the normal distribution applies and the distribution of sample proportions will be N(0.08, 0.016). We use technology to sketch the graph, or we can sketch it by hand, noting that the sample proportions will be centered at the population proportion of 0.08 and roughly 95% of the distribution lies within two standard deviations on either side of the center. This goes between $0.08 \pm 2(0.016)$ or 0.048 to 0.112.

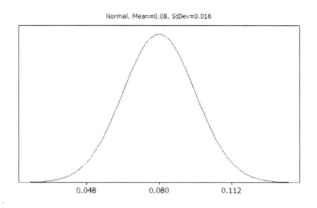

Normal, Mean=0.08, StDev=0.016

0.048   0.080   0.112

6.7 In this case, we have $p = 0.30$ and $n = 500$. The sample proportions will be centered at the population proportion of $p = 0.30$ so will have a mean of 0.30. The standard deviations of the sample proportions is the standard error, which is

$$SE = \sqrt{\frac{p(1-p)}{n}} = \sqrt{\frac{0.30(1-0.30)}{500}} = 0.020$$

6.9 In this case, we have $p = 0.651$ and $n = 50$. The sample proportions will be centered at the population proportion of $p = 0.651$ so will have a mean of 0.651. The standard deviation of the sample proportions is the standard error, which is

$$SE = \sqrt{\frac{p(1-p)}{n}} = \sqrt{\frac{0.651(1-0.651)}{50}} = 0.067$$

6.11 We compute the standard errors using the formula:

$$n = 30 : \quad SE = \sqrt{\frac{p(1-p)}{n}} = \sqrt{\frac{0.4(0.6)}{30}} = 0.089$$

$$n = 200 : \quad SE = \sqrt{\frac{p(1-p)}{n}} = \sqrt{\frac{0.4(0.6)}{200}} = 0.035$$

$$n = 1000 : \quad SE = \sqrt{\frac{p(1-p)}{n}} = \sqrt{\frac{0.4(0.6)}{1000}} = 0.015$$

We see that as the sample size goes up, the standard error goes down. If the standard error goes down, the sample proportions are less spread out from the population proportion, so the accuracy is better.

6.13 In each case, we determine whether $np \geq 10$ and $n(1-p) \geq 10$.

(a) Yes, the conditions apply, since $np = 500(0.1) = 50$ and $n(1-p) = 500(1-0.1) = 450$.

(b) Yes, the conditions apply, since $np = 25(0.5) = 12.5$ and $n(1-p) = 25(1-0.5) = 12.5$.

(c) No, the conditions do not apply, since $np = 30(0.2) = 6 < 10$.

(d) No, the conditions do not apply, since $np = 100(.92) = 92$ but $n(1-p) = 100(1-0.92) = 8 < 10$.

6.15 (a) The graph below shows the sample proportions for 1000 samples of size 50 which were simulated from a population where $p = 0.25$. The distribution is relatively symmetric and bell-shaped so a normal distribution is appropriate. The mean of these simulated sample proportions is 0.252 and the standard error is 0.062. Answers will vary with different simulations but will always be approximately 0.25 and 0.061, respectively.

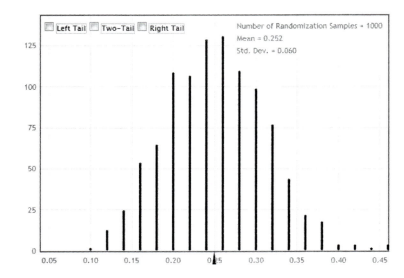

(b) The Central Limit Theorem for proportions says the expected mean of the sample proportions is p, which is 0.25, and the expected standard error is

$$SE = \sqrt{\frac{p(1-p)}{n}} = \sqrt{\frac{0.25(1-0.25)}{50}} = 0.061$$

These are both very close to those obtained with simulations.

6.17 (a) The graph below shows the sample proportions for 1000 samples of size 40 which were simulated from a population where $p = 0.5$. The distribution is relatively symmetric and bell-shaped so a normal distribution is appropriate. The mean of these simulated sample proportions is 0.499 and the standard error is 0.074. Answers will vary with different simulations but will always be approximately 0.5 and 0.079, respectively.

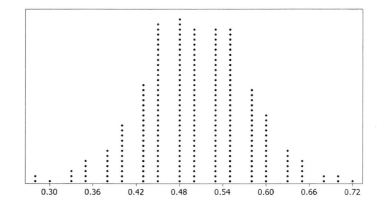

(b) The Central Limit Theorem for proportions says the expected mean of the sample proportions is p, which is 0.5, and the expected standard error is

$$SE = \sqrt{\frac{p(1-p)}{n}} = \sqrt{\frac{0.5(1-0.5)}{40}} = 0.079$$

These are both very close to those obtained with simulations.

6.19 Using *StatKey* or other technology to create a bootstrap distribution, we see for one set of 1000 simulations that $SE = 0.05$. (Answers may vary slightly with other simulations.) Using the formula from the Central Limit Theorem, and using $\hat{p} = 0.52$ as an estimate for p, we have

$$SE = \sqrt{\frac{p(1-p)}{n}} \approx \sqrt{\frac{0.52(1-0.52)}{100}} = 0.050$$

We see that the bootstrap standard error and the formula match very closely.

6.21 Using *StatKey* or other technology to create a bootstrap distribution, we see for one set of 1000 simulations that $SE = 0.068$. (Answers may vary slightly with other simulations.) Using the formula from the Central Limit Theorem, and using $\hat{p} = 0.354$ as an estimate for p, we have

$$SE = \sqrt{\frac{p(1-p)}{n}} \approx \sqrt{\frac{0.354(1-0.354)}{48}} = 0.069$$

We see that the bootstrap standard error and the formula match very closely.

6.23 (a) Since $p = 0.64$ is the proportion of countries in the population that have values for *Energy*, the distribution of sample proportions for samples of 20 countries each will have mean of 0.64 (the population proportion). The standard error of the proportions is $SE = \sqrt{\frac{0.64 \cdot (1-0.64)}{20}} = 0.107$.

(b) We have $np = 20 \cdot 0.64 = 12.8 > 10$, but $n(1-p) = 20(1-.64) = 7.2 < 10$; so we should not expect the proportions from these samples to follow a normal distribution.

(c) If we record the proportion of countries with missing *Energy* values in each sample, the population proportion is 0.36, rather than 0.64. This changes the mean for the distribution of sample proportions to 0.36, but the other answers (for example, $SE = 0.107$) remain the same since we are just switching the roles of p and $1-p$.

6.25 The sample proportions are normally distributed with mean 0.75 and standard deviation $SE = \sqrt{p(1-p)/n} = \sqrt{0.75(0.25)/200} = 0.031$. We find the area above 0.80 in a $N(0.75, 0.031)$ normal curve. We see in the figure that this area is 0.0534. Only about 5.34% of random samples of size 200 will produce a free throw percentage of 80% or higher.

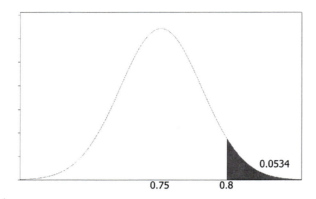

Note that we could also convert the endpoint to a z-score

$$z = \frac{0.80 - 0.75}{0.031} = 1.61$$

and find the area above $z = 1.61$ for a standard N(0,1) density.

Section 6.2 Solutions

6.27 The sample size is large enough to use the normal distribution. For a confidence interval using the normal distribution, we use

$$\text{Sample statistic} \pm z^* \cdot SE.$$

The relevant sample statistic for a confidence interval for a proportion is $\hat{p} = 0.85$. For a 90% confidence interval, we have $z^* = 1.645$, and the standard error is $SE = \sqrt{\hat{p}(1 - \hat{p})/n}$. The confidence interval is

$$
\begin{aligned}
\hat{p} \quad &\pm \quad z^*\sqrt{\frac{\hat{p}(1 - \hat{p})}{n}} \\
0.85 \quad &\pm \quad 1.645 \cdot \sqrt{\frac{0.85(0.15)}{120}} \\
0.85 \quad &\pm \quad 0.054 \\
0.796 \quad &\text{to} \quad 0.904
\end{aligned}
$$

The best estimate for p is 0.85, the margin of error is ± 0.054, and the 90% confidence interval for p is 0.796 to 0.904.

6.29 The sample size is large enough to use the normal distribution. For a confidence interval using the normal distribution, we use

$$\text{Sample statistic} \pm z^* \cdot SE.$$

The relevant sample statistic for a confidence interval for a proportion is $\hat{p} = 0.23$. For a 95% confidence interval, we have $z^* = 1.96$, and the standard error is $SE = \sqrt{\hat{p}(1 - \hat{p})/n}$. The confidence interval is

$$
\begin{aligned}
\hat{p} \quad &\pm \quad z^*\sqrt{\frac{\hat{p}(1 - \hat{p})}{n}} \\
0.23 \quad &\pm \quad 1.96 \cdot \sqrt{\frac{0.23(0.77)}{400}} \\
0.23 \quad &\pm \quad 0.041 \\
0.189 \quad &\text{to} \quad 0.271
\end{aligned}
$$

The best estimate for the proportion of the population in category A is 0.23, the margin of error is ± 0.041, and the 95% confidence interval for the proportion in category A is 0.189 to 0.271.

6.31 The desired margin of error is $ME = 0.01$ and we have $z^* = 2.576$ for 99% confidence. Since we are given no information about the population parameter, we use the conservative estimate $\tilde{p} = 0.5$. We use the formula to find sample size:

$$n = \left(\frac{z^*}{ME}\right)^2 \tilde{p}(1 - \tilde{p}) = \left(\frac{2.576}{0.01}\right)^2 (0.5 \cdot 0.5) = 16{,}589.4.$$

We round up to $n = 16{,}590$. In order to ensure that the margin of error is within the desired $\pm 1\%$, we should use a sample size of 16,590 or higher. This is a very large sample size! The sample size goes up significantly as we aim for greater accuracy and greater confidence in the result.

6.33 The desired margin of error is $ME = 0.02$ and we have $z^* = 1.96$ for 95% confidence. We use the information we have about the sample proportion, so we use $\hat{p} = 0.78$ as our estimate for the population parameter. We use the formula to find sample size:

$$n = \left(\frac{z^*}{ME}\right)^2 \tilde{p}(1 - \tilde{p}) = \left(\frac{1.96}{0.02}\right)^2 (0.78 \cdot 0.22) = 1648.05.$$

We round up to $n = 1649$. In order to ensure that the margin of error is within the desired $\pm 2\%$, we should use a sample size of 1649 or higher.

6.35 The sample size is definitely large enough to use the normal distribution. For a confidence interval using the normal distribution, we use

$$\text{Sample statistic} \pm z^* \cdot SE.$$

The relevant sample statistic for a confidence interval for a proportion is $\hat{p}$, and we have $\hat{p} = 959/1068 = 0.898$. For a 95% confidence interval, we have $z^* = 1.96$, and the standard error is $SE = \sqrt{\hat{p}(1 - \hat{p})/n}$. The confidence interval is

$$
\begin{aligned}
\hat{p} \quad &\pm \quad z^* \sqrt{\frac{\hat{p}(1 - \hat{p})}{n}} \\
0.898 \quad &\pm \quad 1.96 \cdot \sqrt{\frac{0.898(0.102)}{1068}} \\
0.898 \quad &\pm \quad 0.018 \\
0.880 \quad &\text{to} \quad 0.916
\end{aligned}
$$

We are 95% confident that the proportion of times NFL teams punt on a fourth down when the analysis shows they should not punt is between 0.880 and 0.916.

6.37 The sample size is clearly large enough to use the formula based on the normal approximation, since there are well more than 10 responses in each category.

(a) The proportion in the sample who disagreed is $\hat{p} = 1812/2625 = 0.69$ and $z^* = 1.645$ for 90% confidence, so we have

$$0.69 \pm 1.645 \sqrt{\frac{0.69(1 - 0.69)}{2625}} = 0.69 \pm 0.015 = (0.675, 0.705)$$

We are 95% sure that between 67.5% and 70.5% of people would disagree with the statement "There is only one true love for each person."

(b) The proportion in the sample who answered "don't know" is $\hat{p} = 78/2625 = 0.03$ so the 90% confidence interval is

$$0.03 \pm 1.645 \sqrt{\frac{0.03(1 - 0.03)}{2625}} = 0.03 \pm 0.005 = (0.025, 0.035)$$

We are 90% sure that between 2.5% and 3.5% of people would respond with "don't know."

(c) The estimated proportion of people who disagree (which is closer to 0.5) has a larger margin of error.

6.39 The sample sizes ($n = 2255$ US adults or $n = 1787$ Internet users) are both large enough for the proportions in this exercise to follow a normal distribution. In each case we find the confidence interval with

$$\text{Sample statistic} \pm z^* \cdot SE.$$

where $z^* = 1.96$ for 95% confidence and $SE = \sqrt{\hat{p}(1 - \hat{p})/n}$.

(a) For the proportion of US adults who use the Internet regularly, we find $\hat{p} = 1787/2255 = 0.792$. The confidence interval is

$$\hat{p} \quad \pm \quad z^* \sqrt{\frac{\hat{p}(1 - \hat{p})}{n}}$$

$$0.792 \quad \pm \quad 1.96 \cdot \sqrt{\frac{0.792(0.208)}{2255}}$$

$$0.792 \quad \pm \quad 0.017$$

$$0.775 \quad \text{to} \quad 0.809$$

We are 95% confident that the proportion of US adults who use the Internet regularly is between 0.775 and 0.809.

(b) For the proportion of Internet users who use social networking sites, we find $\hat{p} = 1054/1787 = 0.590$. The confidence interval is

$$\hat{p} \quad \pm \quad z^* \sqrt{\frac{\hat{p}(1 - \hat{p})}{n}}$$

$$0.590 \quad \pm \quad 1.96 \cdot \sqrt{\frac{0.590(0.410)}{1787}}$$

$$0.590 \quad \pm \quad 0.023$$

$$0.567 \quad \text{to} \quad 0.613$$

We are 95% confident that the proportion of US adult Internet users who use social networking sites is between 0.567 and 0.613.

(c) For the proportion of US adults who use social networking sites, we find $\hat{p} = 1054/2255 = 0.467$. The confidence interval is

$$\hat{p} \quad \pm \quad z^* \sqrt{\frac{\hat{p}(1 - \hat{p})}{n}}$$

$$0.467 \quad \pm \quad 1.96 \cdot \sqrt{\frac{0.467(0.533)}{2255}}$$

$$0.467 \quad \pm \quad 0.021$$

$$0.446 \quad \text{to} \quad 0.488$$

We are 95% confident that the proportion of US adults who use a social networking site between 0.446 and 0.488. Since 0.5 is not within this range of plausible values for the population proportion, it is not plausible to estimate that 50% of US adult Internet users use a social networking site. (The percentage is increasing rapidly though, and doubled in the three years from 2008 to 2011. By the time you are reading this book, the percentage will probably be over 50%.)

6.41 (a) The sample is the 27,000 people included in the survey. The population is all consumers with Internet access.

(b) The sample size is definitely large enough to use the normal distribution. For a confidence interval using the normal distribution, we use

$$\text{Sample statistic} \pm z^* \cdot SE.$$

The relevant sample statistic for a confidence interval for a proportion is $\hat{p} = 0.61$. For a 99% confidence interval, we have $z^* = 2.576$, and the standard error is $SE = \sqrt{\hat{p}(1-\hat{p})/n}$. The confidence interval is

$$\hat{p} \quad \pm \quad z^* \sqrt{\frac{\hat{p}(1-\hat{p})}{n}}$$

$$0.61 \quad \pm \quad 2.576 \cdot \sqrt{\frac{0.61(0.39)}{27{,}000}}$$

$$0.61 \quad \pm \quad 0.008$$

$$0.602 \quad \text{to} \quad 0.618$$

We are 99% confident that the proportion of consumers worldwide who purchased more store brands during the economic downturn is between 0.602 and 0.618. The margin of error is very small because the sample size is so large.

(c) The only change for this part is that we now have $\hat{p} = 0.91$. The confidence interval is

$$\hat{p} \quad \pm \quad z^* \sqrt{\frac{\hat{p}(1-\hat{p})}{n}}$$

$$0.91 \quad \pm \quad 2.576 \cdot \sqrt{\frac{0.91(0.09)}{27{,}000}}$$

$$0.91 \quad \pm \quad 0.004$$

$$0.906 \quad \text{to} \quad 0.914$$

We are 99% confident that the proportion of consumers worldwide who plan to continue to purchase the same number of store brands is between 0.906 and 0.914. The margin of error is even smaller as the sample proportion gets farther away from 0.5.

6.43 For 95% confidence, we have $z^* = 1.96$, so the margin of error for estimating the proportion of Democrats is

$$ME = z^* \cdot \sqrt{\frac{\hat{p}(1-\hat{p})}{n}} = 1.96 \cdot \sqrt{\frac{0.351(0.649)}{15000}} = 0.0076.$$

The margin of error for estimating the proportion of Republicans is

$$ME = z^* \cdot \sqrt{\frac{\hat{p}(1-\hat{p})}{n}} = 1.96 \cdot \sqrt{\frac{0.321(0.679)}{15000}} = 0.0075.$$

The proportion of Democrats might be as low as $0.351 - 0.0076 = 0.3434$ while the proportion of Republicans might be as high as $0.321 + 0.0075 = 0.3285$. Even at the extremes of the confidence intervals, the proportion of Democrats is still higher. Thus we can feel comfortable concluding that more American adults self-identified as Democrats than as Republicans in February 2010.

6.45 Using *StatKey* or other technology we create a bootstrap distribution with at least 1000 simulated proportions. To find a 95% confidence interval we find the endpoints that contain 95% of the simulated proportions. For one set of 1000 bootstrap proportions we find that a 95% confidence interval for the proportion of orange Reese's Pieces goes from 0.40 to 0.56.

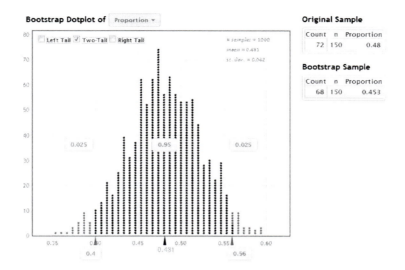

Using the normal distribution and the formula for standard error, we have

$$0.48 \pm 1.96 \cdot \sqrt{\frac{0.48(0.52)}{150}} = 0.48 \pm 0.080 = (0.40, 0.56).$$

In this case, the two methods give exactly the same interval.

6.47 We have $ME = 0.03$ for the margin of error. Since we are given no information about the population proportion, we use the conservative estimate $\tilde{p} = 0.5$.

For 99% confidence, we use $z^* = 2.576$. We have:

$$n = \left(\frac{z^*}{ME}\right)^2 \tilde{p}(1 - \tilde{p}) = \left(\frac{2.576}{0.03}\right)^2 (0.5 \cdot 0.5) = 1843.3$$

We round up to $n = 1844$.

For 95% confidence, we use $z^* = 1.96$. We have:

$$n = \left(\frac{z^*}{ME}\right)^2 \tilde{p}(1 - \tilde{p}) = \left(\frac{1.96}{0.03}\right)^2 (0.5 \cdot 0.5) = 1067.1$$

We round up to $n = 1068$.

For 90% confidence, we use $z^* = 1.645$. We have:

$$n = \left(\frac{z^*}{ME}\right)^2 \tilde{p}(1 - \tilde{p}) = \left(\frac{1.645}{0.03}\right)^2 (0.5 \cdot 0.5) = 751.7$$

We round up to $n = 752$.

We see that the sample size goes up as the level of confidence we want in the result goes up. Or, put another way, a larger sample size gives a higher level of confidence in the accuracy of the estimate.

6.49   (a) The sample size is definitely large enough to use the normal distribution. The relevant sample statistic is $\hat{p} = 0.32$ and, for a 95% confidence interval, we use $z^* = 1.96$. The confidence interval is

$$\hat{p} \quad \pm \quad z^* \sqrt{\frac{\hat{p}(1 - \hat{p})}{n}}$$

$$0.32 \quad \pm \quad 1.96 \cdot \sqrt{\frac{0.32(1 - 0.32)}{1000}}$$

$$0.32 \quad \pm \quad 0.029$$

$$0.291 \quad \text{to} \quad 0.349$$

We are 95% confident that the proportion of US adults who favor a tax on soda and junk food is between 0.291 and 0.349.

(b) The margin of error is 2.9%.

(c) Since the margin of error is about $\pm 3\%$ with a sample size of $n = 1000$, we'll definitely need a sample size larger than 1000 to get the margin of error down to $\pm 1\%$. To see how much larger, we use the formula for determining sample size. The margin of error we desire is $ME = 0.01$, and for 95% confidence we use $z^* = 1.96$. We can use the sample statistic $\hat{p} = 0.32$ as our best estimate for p. We have:

$$n \quad = \quad \left(\frac{z^*}{ME}\right)^2 \tilde{p}(1 - \tilde{p})$$

$$= \quad \left(\frac{1.96}{0.01}\right)^2 0.32(1 - 0.32)$$

$$= \quad 8359.3.$$

We round up, so we would need to include 8,360 people in the survey in order to get the margin of error down to within $\pm 1\%$.

6.51 Since we have no reasonable estimate for the proportion, we use $\tilde{p} = 0.5$. For 98% confidence, use $z^* = 2.326$. The required sample size is

$$n = \left(\frac{2.326}{0.04}\right)^2 0.5(1 - 0.5) = 845.4$$

We round up to show a sample of at least 846 individuals is needed to estimate the proportion who would consider buying a sunscreen pill to within 4% with 98% confidence.

6.53 The data in **CommuteAtlanta** contains 254 males and 246 females among the 500 commuters. Thus the sample proportion of males is $\hat{p} = 254/500 = 0.508$. Since there are more than 10 males and 10 females in the sample, we may use the normal approximation to construct a confidence interval for the proportion of males. For 95% confidence the standard normal endpoints are $z^* = 1.96$, so we compute the interval with

$$0.508 \pm 1.96 \sqrt{\frac{0.508(1 - 0.508)}{500}} = 0.508 \pm 0.044 = (0.464, 0.552)$$

We are 95% sure that somewhere between 46.4% and 55.2% of Atlanta commuters are male.

6.55 The sample proportion who prefer a playoff system is $\hat{p} = 0.63$, the margin of error is $ME = 0.031$, and $z^* = 1.96$ for 95% confidence. We can "work backwards" to determine the sample size:

$$n = \left(\frac{1.96}{0.031}\right)^2 0.63(1 - 0.63) = 931.8$$

So, about 932 college football fans were sampled.

Section 6.3 Solutions

6.57 Since $np_0 = 200(0.3) = 60$ and $n(1 - p_0) = 200(0.7) = 140$, the sample size is large enough to use the normal distribution. In general, the standardized test statistic is

$$z = \frac{\text{Sample Statistic} - \text{Null Parameter}}{SE}.$$

In this test for a proportion, the sample statistic is $\hat{p} = 0.21$ and the parameter from the null hypothesis is $p_0 = 0.3$. The standard error is $SE = \sqrt{p_0(1 - p_0)/n}$. The standardized test statistic is

$$z = \frac{\hat{p} - p_0}{\sqrt{\frac{p_0(1-p_0)}{n}}} = \frac{0.21 - 0.3}{\sqrt{\frac{0.3(0.7)}{200}}} = -2.78.$$

This is a lower-tail test, so the p-value is the area below -2.78 in a standard normal distribution. Using technology or a table, we see that the p-value is 0.0027. This p-value is very small so we find strong evidence to support the alternative hypothesis that $p < 0.3$.

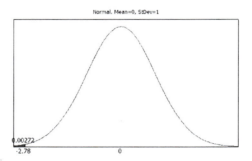

6.59 Since $np_0 = 50(0.8) = 40$ and $n(1 - p_0) = 50(0.2) = 10$, the sample size is (just barely) large enough to use the normal distribution. In general, the standardized test statistic is

$$z = \frac{\text{Sample Statistic} - \text{Null Parameter}}{SE}.$$

In this test for a proportion, the sample statistic is $\hat{p} = 0.88$ and the parameter from the null hypothesis is $p_0 = 0.8$. The standard error is $SE = \sqrt{p_0(1 - p_0)/n}$. The standardized test statistic is

$$z = \frac{\hat{p} - p_0}{\sqrt{\frac{p_0(1-p_0)}{n}}} = \frac{0.88 - 0.80}{\sqrt{\frac{0.8(0.2)}{50}}} = 1.41.$$

This is an upper-tail test, so the p-value is the area above 1.41 in a standard normal distribution. Using technology or a table, we see that the p-value is 0.079. This p-value is larger than the significance level of 5%, so we do not find sufficient evidence to support the alternative hypothesis that $p > 0.8$.

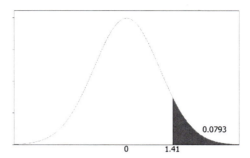

6.61 Since $np_0 = 1000(0.2) = 200$ and $n(1 - p_0) = 1000(0.8) = 800$, the sample size is large enough to use the normal distribution. In general, the standardized test statistic is

$$z = \frac{\text{Sample Statistic} - \text{Null Parameter}}{SE}.$$

In this test for a proportion, the sample statistic is $\hat{p} = 0.26$ and the parameter from the null hypothesis is $p_0 = 0.2$. The standard error is $SE = \sqrt{p_0(1 - p_0)/n}$. The standardized test statistic is

$$z = \frac{\hat{p} - p_0}{\sqrt{\frac{p_0(1-p_0)}{n}}} = \frac{0.26 - 0.2}{\sqrt{\frac{0.2(0.8)}{1000}}} = 4.74.$$

This is a two-tail test, so the p-value is two times the area above 4.74 in a standard normal distribution. The area beyond 4.74 is essentially zero, so the p-value is essentially zero. The p-value is very small so we have strong evidence to support the alternative hypothesis that $p \neq 0.2$.

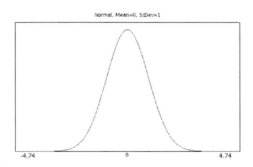

6.63 If we let p represent the proportion of US adults who believe in ghosts, the hypotheses are:

$$H_0: \quad p = 0.25$$
$$H_a: \quad p > 0.25$$

The sample size of 1000 is clearly large enough for us to use the normal distribution. The test statistic is:

$$z = \frac{\text{Sample statistic} - \text{Null parameter}}{SE} = \frac{\hat{p} - p_0}{\sqrt{\frac{p_0(1-p_0)}{n}}} = \frac{0.31 - 0.25}{\sqrt{\frac{0.25(0.75)}{1000}}} = 4.38.$$

This is an upper-tail test, so the p-value is the area above 4.38 in a standard normal distribution. We know that 4.38 (more than four standard deviations above the mean) is going to be way out in the tail of the distribution, and the p-value is essentially zero. There is very strong evidence that more than 1 in 4 US adults believes in ghosts.

6.65 We are conducting a hypothesis test for a proportion p, where p is the proportion of all US adults who know most or all of their neighbors. We are testing to see if there is evidence that $p > 0.5$, so we have

$$H_0: \quad p = 0.5$$
$$H_a: \quad p > 0.5$$

This is a one-tail test since we are specifically testing to see if the proportion is greater than 0.5. The sample proportion is $\hat{p} = 0.51$ and the null proportion is $p_0 = 0.5$. The sample size is $n = 2,255$ so the test statistic is:

$$z = \frac{\text{Sample statistic} - \text{Null parameter}}{SE} = \frac{\hat{p} - p_0}{\sqrt{\frac{p_0(1-p_0)}{n}}} = \frac{0.51 - 0.5}{\sqrt{\frac{0.5(0.5)}{2255}}} = 0.95.$$

This is a one-tail test, so the p-value is the area above 0.95 in the standard normal distribution. We find a p-value of 0.171. This p-value is larger than even a 10% significance level, so we do not have sufficient evidence to conclude that the proportion of US adults who know their neighbors is larger than 0.5.

6.67 We test $H_0: p = 1/3$ vs $H_a: p \neq 1/3$ where p is the proportion of all female students who choose the Olympic gold medal. For the sample of $n = 169$ female students we have $\hat{p} = 73/169 = 0.432$. We compute a standardized test statistic.

$$z = \frac{0.432 - 1/3}{\sqrt{\frac{1/3(1-1/3)}{169}}} = 2.72$$

Checking that $169 \cdot 1/3 = 56.3$ and $169 \cdot (1 - 1/3) = 112.7$ are both bigger than 10, we use the standard normal curve to find the p-value. Using technology or a table, the area to the right beyond 2.72 is 0.003 and we double this value since it's a two-tailed alternative to get p-value $= 0.006$. This is a small p-value, so we have strong evidence that the proportion of female students who choose the Olympic gold medal differs from one third. Thus it would appear that the three awards are not equally popular among female students.

6.69 Let p be the proportion of times Team B will beat Team A. We are testing $H_0: p = 0.5$ vs $H_a: p \neq 0.5$. The sample proportion is $\hat{p} = 24/40 = 0.6$. The test statistic is

$$z = \frac{0.6 - 0.5}{\sqrt{\frac{0.5(1-0.5)}{40}}} = 1.26$$

The area to the right of 1.26 on the standard normal distribution is 0.104, so the p-value is $2(0.104) = 0.208$. There is not convincing evidence that one team is better than the other. (We arrive at the same conclusion if we let p be the proportion of times that Team A wins.)

6.71 We use technology to determine that the number of regular vitamin users in the sample is 122, so the sample proportion is $\hat{p} = 122/315 = 0.387$. The hypotheses are:

$$H_0: \quad p = 0.35$$
$$H_a: \quad p \neq 0.35$$

The test statistic is:

$$z = \frac{\text{Sample statistic} - \text{Null parameter}}{SE} = \frac{\hat{p} - p_0}{\sqrt{\frac{p_0(1-p_0)}{n}}} = \frac{0.387 - 0.35}{\sqrt{\frac{0.35(0.65)}{315}}} = 1.38.$$

This is a two-tail test, so the p-value is twice the area above 1.38 in a standard normal distribution. We see that the p-value is $2(0.084) = 0.168$. This p-value is larger than any reasonable significance level, so we do not reject H_0. We do not find evidence that the proportion of people who regularly take a vitamin pill is not 35%.

Section 6.4 Solutions

6.73 (a) The sample means will have a mean of 6 and a standard error of

$$SE = \frac{\sigma}{\sqrt{n}} = \frac{2}{\sqrt{10}} = 0.632.$$

(b) Since $n < 30$, the normal distribution does not apply.

6.75 (a) The sample means will have a mean of 60 and a standard error of

$$SE = \frac{\sigma}{\sqrt{n}} = \frac{32}{\sqrt{75}} = 3.695.$$

(b) Since $n \geq 30$, the normal distribution applies and the distribution of sample means will be N(60,3.695). We use technology to sketch the graph, or we can sketch it by hand, noting that the sample means will be centered at the population mean of 60 and roughly 95% of the distribution lies within two standard deviations on either side of the center. This goes between $60 \pm 2(3.695)$ or 52.61 to 67.39.

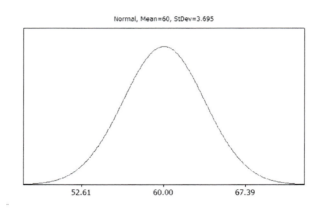

6.77 We use a t-distribution with df = 17. Using technology, we see that the values with 1% beyond them in each tail are ± 2.57.

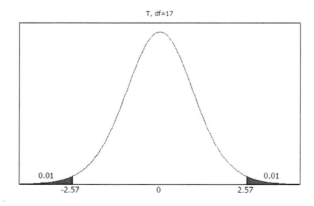

6.79 We use a t-distribution with df = 39. Using technology, we see that the values with 0.005 beyond them in each tail are ±2.71.

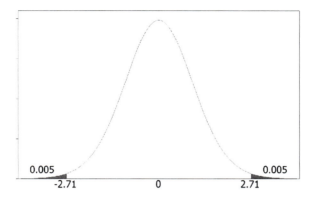

6.81 We use a t-distribution with df = 7. Using technology, we see that the area above 1.5 is 0.0886. (On a paper table, we may only be able to specify that the area is between 0.05 and 0.10.)

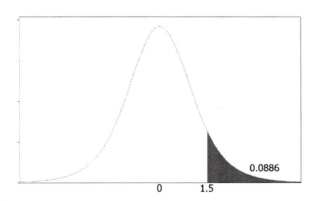

6.83 We use a t-distribution with df = 49. Using technology, we see that the area below -3.2 is 0.0012.

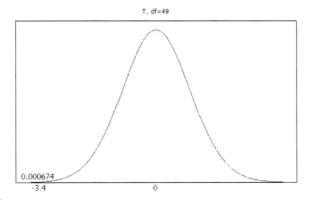

6.85 (a) The mean of the distribution is 230 minutes. The standard deviation of the distribution of sample means is the standard error:

$$SE = \frac{\sigma}{\sqrt{n}} = \frac{38}{\sqrt{10}} = 12.0 \text{ minutes.}$$

(b) The mean of the distribution is 230 minutes. The standard deviation of the distribution of sample means is the standard error:

$$SE = \frac{\sigma}{\sqrt{n}} = \frac{38}{\sqrt{100}} = 3.8 \text{ minutes.}$$

(c) The mean of the distribution is 230 minutes. The standard deviation of the distribution of sample means is the standard error:

$$SE = \frac{\sigma}{\sqrt{n}} = \frac{38}{\sqrt{1000}} = 1.2 \text{ minutes.}$$

Notice that as the sample size goes up, the standard error of the sample means goes down.

6.87 The means of the samples will be clustered around the population mean of $\mu = 516$ so the distribution will be centered at a mean of 516. The standard deviation of the distribution is the standard error:

$$SE = \frac{\sigma}{\sqrt{n}} = \frac{116}{\sqrt{100}} = 11.6$$

6.89 (a) i. The mean is 501 and the standard deviation is 112. Taking samples of size 1 is the same as the original distribution.

ii. The mean is 501 and the standard deviation is $SE = \sigma/\sqrt{n} = 112/\sqrt{10} = 35.42$.

iii. The mean is 501 and the standard deviation is $SE = \sigma/\sqrt{n} = 112/\sqrt{100} = 11.2$.

iv. The mean is 501 and the standard deviation is $SE = \sigma/\sqrt{n} = 112/\sqrt{1000} = 3.54$.

(b) The center of the distribution is not affected by the sample size, but the variability of the sample means goes down as the sample size increases.

6.91 We compute the standard errors using the formula:

$$\sigma = 5 : \quad SE = \frac{\sigma}{\sqrt{n}} = \frac{5}{\sqrt{100}} = 0.5.$$

$$\sigma = 25 : \quad SE = \frac{\sigma}{\sqrt{n}} = \frac{25}{\sqrt{100}} = 2.5.$$

$$\sigma = 75 : \quad SE = \frac{\sigma}{\sqrt{n}} = \frac{75}{\sqrt{100}} = 7.5.$$

If the population standard deviation is larger, the standard error of the sample means will also be larger. This make sense: if the data are more spread out, the sample means (assuming the same sample size) will be more spread out also.

6.93 The t-distribution is appropriate if the sample size is large ($n \geq 30$) or if the underlying distribution appears to be relatively normal. We have concerns about the t-distribution only for small sample sizes and heavy skewness or outliers. In this case, the sample size is large enough ($n = 75$) that we can feel comfortable using the t-distribution, despite the clear skewness in the data. The t-distribution is appropriate. For the degrees of freedom *df* and estimated standard error *SE*, we have:

$$df = n - 1 = 75 - 1 = 74, \quad \text{and} \quad SE = \frac{s}{\sqrt{n}} = \frac{10.1}{\sqrt{75}} = 1.17$$

6.95 The t-distribution is appropriate if the sample size is large ($n \geq 30$) or if the underlying distribution appears to be relatively normal. We have concerns about the t-distribution only for small sample sizes and heavy skewness or outliers. In this case, the sample size is small ($n = 12$) and the data is heavily skewed with some apparent outliers. It would not be appropriate to use the t-distribution in this case. We might try analyzing the data using simulation methods such as a bootstrap or randomization distribution.

6.97 Using *StatKey* or other technology to create a bootstrap distribution, we see for one set of 1000 simulations that $SE \approx 0.92$. (Answers may vary slightly with other simulations.) Using the formula from the Central Limit Theorem, and using $s = 20.72$ as an estimate for σ, we have

$$SE = \frac{s}{\sqrt{n}} = \frac{20.72}{\sqrt{500}} = 0.93.$$

We see that the bootstrap standard error and the formula match very closely.

6.99 Using *StatKey* or other technology to create a bootstrap distribution, we see for one set of 1000 simulations that $SE \approx 0.11$. (Answers may vary slightly with other simulations.) Using the formula from the Central Limit Theorem, and using $s = 0.765$ as an estimate for σ, we have

$$SE = \frac{s}{\sqrt{n}} = \frac{0.765}{\sqrt{50}} = 0.108.$$

We see that the bootstrap standard error and the formula match very closely.

6.101 Definitely not NY since the data are so skewed. Maybe NJ and probably not PA.

6.103 The underlying distribution is normal, so the distribution of sample means is normal for *any* sample size. The standard error of the sample mean will change as n changes.

(a)
    i. If the sample size is 1, the distribution of sample means is the same as the distribution of scores, which is N(501,112). For this distribution, the area in the tail beyond 525 is 0.415. Approximately 41.5% of the scores will be greater than or equal to 525.

    ii. If the sample size is 10, the distribution of sample means is N(501,35.4), since $SE = \sigma/\sqrt{n} = 112/\sqrt{10} = 35.4$. For this distribution, the area in the tail beyond 525 is 0.249. Approximately 24.9% of the means from random samples of size 10 will be greater than or equal to 525.

    iii. If the sample size is 100, the distribution of sample means is N(501,11.2), since $SE = \sigma/\sqrt{n} = 112/\sqrt{100} = 11.2$. For this distribution, the area in the tail beyond 525 is 0.0161. Only about 1.6% of the means from random samples of size 100 will be greater than or equal to 525.

    iv. If the sample size is 1000, the distribution of sample means is N(501,3.54), since $SE = \sigma/\sqrt{n} = 112/\sqrt{100} = 3.54$. For this distribution, the area in the tail beyond 525 is essentially zero. None of the means from random samples of size 1000 will be greater than or equal to 525.

(b) As the sample size increases, the sample means get closer to the population mean (if the samples are random samples!) and the likelihood of being far from the population mean goes down. Indeed, although 41.5% of the individual scores are 525 or higher, it would be extremely unusual to see a sample mean that large from a random sample of size 1000, and even quite unusual to see such a large sample mean from a random sample of size 100.

Section 6.5 Solutions

6.105 For a confidence interval for μ using the t-distribution, we use

$$\overline{x} \pm t^* \cdot \frac{s}{\sqrt{n}}$$

We use a t-distribution with df = 29, so for a 95% confidence interval, we have $t^* = 2.05$. The confidence interval is

$$
\begin{aligned}
12.7 \quad &\pm \quad 2.05 \cdot \frac{5.6}{\sqrt{30}} \\
12.7 \quad &\pm \quad 2.10 \\
10.6 \quad &\text{to} \quad 14.8
\end{aligned}
$$

The best estimate for μ is $\overline{x} = 12.7$, the margin of error is ± 2.10, and the 95% confidence interval for μ is 10.6 to 14.8. We are 95% confident that the mean of the entire population is between 10.6 and 14.8.

6.107 For a confidence interval for μ using the t-distribution, we use

$$\overline{x} \pm t^* \cdot \frac{s}{\sqrt{n}}$$

We use a t-distribution with df = 99, so for a 90% confidence interval, we have $t^* = 1.66$. (Since the sample size is so large, the t distribution value is almost identical to the standard normal z value.) The confidence interval is

$$
\begin{aligned}
3.1 \quad &\pm \quad 1.66 \cdot \frac{0.4}{\sqrt{100}} \\
3.1 \quad &\pm \quad 0.066 \\
3.034 \quad &\text{to} \quad 3.166
\end{aligned}
$$

The best estimate for μ is $\overline{x} = 3.1$, the margin of error is ± 0.066, and the 90% confidence interval for μ is 3.034 to 3.166. We are 90% confident that the mean of the entire population is between 3.034 and 3.166.

6.109 For a confidence interval for μ using the t-distribution, we use

$$\overline{x} \pm t^* \cdot \frac{s}{\sqrt{n}}$$

We use a t-distribution with df = 9, so for a 99% confidence interval, we have $t^* = 3.25$. The confidence interval is

$$
\begin{aligned}
46.1 \quad &\pm \quad 3.25 \cdot \frac{12.5}{\sqrt{10}} \\
46.1 \quad &\pm \quad 12.85 \\
33.25 \quad &\text{to} \quad 58.95
\end{aligned}
$$

The best estimate for μ is $\overline{x} = 46.1$, the margin of error is ± 12.85, and the 99% confidence interval for μ is 33.25 to 58.95. We are 99% confident that the mean of the entire population is between 33.35 and 58.95.

6.111 The desired margin of error is $ME = 5$ and we have $z^* = 1.96$ for 95% confidence. We use $\tilde{\sigma} = 18$ to approximate the standard deviation. We use the formula to find sample size:

$$n = \left(\frac{z^* \cdot \tilde{\sigma}}{ME} \right)^2 = \left(\frac{1.96 \cdot 18}{5} \right)^2 = 49.8$$

We round up to $n = 50$. In order to ensure that the margin of error is within the desired ± 5 units, we should use a sample size of 50 or higher.

6.113 The desired margin of error is $ME = 0.5$ and we have $z^* = 1.645$ for 90% confidence. We use $\tilde{\sigma} = 25$ to approximate the standard deviation. We use the formula to find sample size:

$$n = \left(\frac{z^* \cdot \tilde{\sigma}}{ME} \right)^2 = \left(\frac{1.645 \cdot 25}{0.5} \right)^2 = 6765.1$$

We round up to $n = 6766$. In order to ensure that the margin of error is within the desired ± 0.5 units, we would need to use a sample size of 6,766 or higher.

6.115 We use a t-distribution with df $= 360$, so for a 99% confidence interval, we have $t^* = 2.59$. The confidence interval is

$$\begin{aligned}
\overline{x} \quad &\pm \quad t^* \cdot \frac{s}{\sqrt{n}} \\
6.504 \quad &\pm \quad 2.59 \cdot \frac{5.584}{\sqrt{361}} \\
6.504 \quad &\pm \quad 0.761 \\
5.743 \quad &\text{to} \quad 7.265
\end{aligned}$$

We are 99% confident that the average number of hours of television watched per week by students who take the introductory statistics course at this university is between 5.743 and 7.265.

6.117 We use a t-distribution with df $= 49$, so for a 95% confidence interval, we have $t^* = 2.01$. The confidence interval is

$$\begin{aligned}
\overline{x} \quad &\pm \quad t^* \cdot \frac{s}{\sqrt{n}} \\
3.1 \quad &\pm \quad 2.01 \cdot \frac{0.72}{\sqrt{50}} \\
3.1 \quad &\pm \quad 0.20 \\
2.9 \quad &\text{to} \quad 3.3
\end{aligned}$$

The best estimate for the length of gribbles is 3.1 mm, with a margin of error for our estimate of ± 0.2. The 95% confidence interval is 2.9 to 3.3, and we are 95% confident that the average length of all gribbles is between 2.9 and 3.3 mm. We need to assume that the sample of gribbles is a random sample, or at least a representative sample.

6.119 The parameter we are estimating is mean weight gain over the time of the study of all mice living with dim light at night. To find a confidence interval for a mean using the t-distribution, we use

$$\text{Sample statistic } \pm t^* \cdot SE.$$

The relevant sample statistic is $\bar{x} = 7.9$. For a 90% confidence interval with $10 - 1 = 9$ degrees of freedom, we use $t^* = 1.83$ and the standard error is $SE = s/\sqrt{n}$. The confidence interval is

$$
\begin{array}{ccc}
\bar{x} & \pm & t^* \cdot \dfrac{s}{\sqrt{n}} \\[2mm]
7.9 & \pm & 1.83 \cdot \dfrac{3.0}{\sqrt{10}} \\[2mm]
7.9 & \pm & 1.74 \\[1mm]
6.16 & \text{to} & 9.64
\end{array}
$$

We are 90% confident that the mean weight gain of mice living with dim light at night will be between 6.16 and 9.64 grams.

6.121   (a) We see that $n = 157$, $\bar{x} = 3.849$, and $s = 2.421$.

(b) We have

$$
SE = \frac{s}{\sqrt{n}} = \frac{2.421}{\sqrt{157}} = 0.193
$$

This is the same as the value given in the computer output.

(c) For a 95% confidence interval with degrees of freedom 156, we use $t^* = 1.98$. The confidence interval is

$$
\begin{array}{ccc}
\bar{x} & \pm & t^* \cdot \dfrac{s}{\sqrt{n}} \\[2mm]
3.849 & \pm & 1.98 \cdot \dfrac{2.421}{\sqrt{157}} \\[2mm]
3.849 & \pm & 0.383 \\[1mm]
3.466 & \text{to} & 4.232
\end{array}
$$

A 95% confidence interval is $3.466 to $4.232.

(d) Up to two decimal places, the confidence interval we found is the same as the one given in the computer output. The small differences are probably due to round-off error in estimating the t^* value.

(e) We are 95% confident that the average tip given at this restaurant is between $3.47 and $4.23.

6.123 We use *StatKey* or other technology to create a bootstrap distribution with at least 1000 simulated means from samples of the Atlanta commute distances. To find a 95% confidence interval we find the endpoints that contain 95% of the simulated means. For one set of 1000 bootstrap means shown below we find that a 95% confidence interval for mean Atlanta commute distance goes from 16.92 to 19.34 miles.

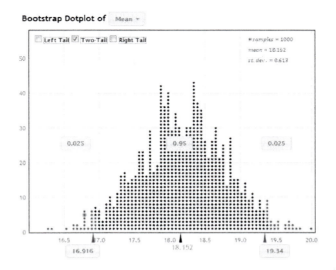

For a 95% confidence interval with $df = 499$, we have $t^* = 1.965$. Using the t-distribution and the formula for standard error, we have

$$18.156 \pm 1.965 \cdot \frac{13.798}{\sqrt{500}} = 18.156 \pm 1.21 = (16.95, 19.37)$$

The two methods give very similar intervals.

6.125   (a) We find that $\overline{x} = 12.20$ penalty minutes per game with $s = 2.25$.

(b) We use a t-distribution with df $= 29$, so for a 95% confidence interval, we have $t^* = 2.05$. The confidence interval is

$$\overline{x} \quad \pm \quad t^* \cdot \frac{s}{\sqrt{n}}$$
$$12.20 \quad \pm \quad 2.05 \cdot \frac{2.25}{\sqrt{30}}$$
$$12.20 \quad \pm \quad 0.84$$
$$11.36 \quad \text{to} \quad 13.04$$

We are 95% confident that the average number of penalty minutes per game for NHL teams is between 11.36 and 13.04.

(c) It might be reasonable to generalize to the broader population of all NHL teams in all years if we think the 2010-2011 teams are representative of all other years. It is probably not appropriate, however, since there are many things that can change year to year. For example, the referees might have been particularly harsh (or lenient) in 2010-2011, or there might be a rule change that caused more or fewer penalties. There are many possible ways in which this sample could be biased.

6.127 With a sample size of 2,006, we can show that the margin of error for the original study is 0.08. If we want the smaller 0.05 margin of error, we will need a larger sample size. How much larger? We find the needed sample size using

$$n = \left(\frac{z^* \cdot \tilde{\sigma}}{ME} \right)^2$$

For 99% confidence, we use $z^* = 2.576$ and we use the standard deviation 1.4 from our earlier sample as our estimated standard deviation $\tilde{\sigma}$. The desired margin of error is $ME = 0.05$. Using the formula, we have

$$n = \left(\frac{z^* \cdot \tilde{\sigma}}{ME}\right)^2 = \left(\frac{2.576 \cdot 1.4}{0.05}\right)^2 = 5202.45$$

Since sample size must be an integer, we round up and recommend a sample of size 5,203 for a margin of error of 0.05, with 99% confidence.

6.129   (a) For a confidence interval for μ using the t-distribution, we use

$$\overline{x} \pm t^* \cdot \frac{s}{\sqrt{n}}$$

Since the sample size is so large, we can use either a t-distribution or a standard normal distribution. Using a t-distribution with df $= 89$, for a 99% confidence interval, we have $t^* = 2.63$. The confidence interval is

$$
\begin{array}{rcl}
18.3 & \pm & 2.63 \cdot \dfrac{8.2}{\sqrt{90}} \\
18.3 & \pm & 2.27 \\
16.03 & \text{to} & 20.57
\end{array}
$$

We are 99% confident that the mean number of polyester microfibers found on the world's beaches is between 16.03 and 20.57 particles per 250 mL of sediment.

(b) The margin of error is ± 2.27 microfibers per liter.

(c) Since the margin of error is ± 2.27 with a sample size of $n = 90$, we'll need a sample size larger than 90 to get the margin of error down to ± 1. To see how much larger, we use the formula for determining sample size. The margin of error we desire is $ME = 1$, and for 99% confidence we use $z^* = 2.576$. We can use the sample statistic $s = 8.2$ as our best estimate for σ. We have:

$$n = \left(\frac{z^* \cdot \tilde{\sigma}}{ME}\right)^2 = \left(\frac{2.576 \cdot 8.2}{1}\right)^2 = 446.2$$

We round up to $n = 447$. We would need to obtain data from 447 beach sites to get the margin of error down to ± 1 particle per 250 mL of sediment.

6.131 We have $ME = 3$ for the margin of error, and $\tilde{\sigma} = 30$ for our estimate of the standard deviation.

For 99% confidence, we use $z^* = 2.576$ to give:

$$n = \left(\frac{z^* \cdot \tilde{\sigma}}{ME}\right)^2 = \left(\frac{2.576 \cdot 30}{3}\right)^2 = 663.6$$

We round up to $n = 664$.

For 95% confidence, we use $z^* = 1.96$ to give:

$$n = \left(\frac{z^* \cdot \tilde{\sigma}}{ME}\right)^2 = \left(\frac{1.96 \cdot 30}{3}\right)^2 = 384.2$$

We round up to $n = 385$.

For 90% confidence, we use $z^* = 1.645$ to give:

$$n = \left(\frac{z^* \cdot \tilde{\sigma}}{ME}\right)^2 = \left(\frac{1.645 \cdot 30}{3}\right)^2 = 270.6$$

We round up to $n = 271$.

We see that the sample size goes up as the level of confidence we want in the result goes up. Or, put another way, if we want greater confidence that the interval for a specific margin of error captures the population mean, we need to use a larger sample size.

6.133 Using any statistics package, we see that a 95% confidence interval 16.62 ± 0.69 or is 15.93 to 17.31. At this restaurant, the average percent added for a tip on a bill is between 15.93% and 17.31%.

Section 6.6 Solutions

6.135 In general, the standardized test statistic is

$$\frac{\text{Sample Statistic} - \text{Null Parameter}}{SE}.$$

In this test for a mean, the sample statistic is $\overline{x} = 17.2$ and the parameter from the null hypothesis is $\mu_0 = 15$. The standard error is $SE = s/\sqrt{n}$. The standardized test statistic is

$$t = \frac{\overline{x} - \mu_0}{s/\sqrt{n}} = \frac{17.2 - 15}{6.4/\sqrt{40}} = 2.17.$$

This is an upper-tail test, so the p-value is the area above 2.17 in a t-distribution with df $= 39$. We see that the p-value is 0.0181. This p-value is relatively small, so at a 5% level, we do find evidence to support the alternative hypothesis that $\mu > 15$.

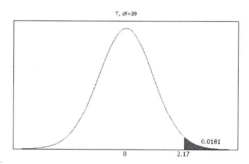

6.137 In general, the standardized test statistic is

$$\frac{\text{Sample Statistic} - \text{Null Parameter}}{SE}.$$

In this test for a mean, the sample statistic is $\overline{x} = 112.3$ and the parameter from the null hypothesis is $\mu_0 = 120$. The standard error is $SE = s/\sqrt{n}$. The standardized test statistic is

$$t = \frac{\overline{x} - \mu_0}{s/\sqrt{n}} = \frac{112.3 - 120}{18.4/\sqrt{100}} = -4.18.$$

This is a lower-tail test, so the p-value is the area below -4.18 in a t-distribution with df $= 99$. We see that the p-value is 0.00003, or essentially zero. This p-value is very small, and below any reasonable significance level. There is strong evidence to support the alternative hypothesis that $\mu < 100$.

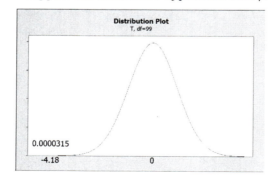

6.139 This sample size is quite small, but we are told that the underlying distribution is approximately normal so we can proceed with the t-test. In general, the standardized test statistic is

$$\frac{\text{Sample Statistic} - \text{Null Parameter}}{SE}.$$

In this test for a mean, the sample statistic is $\bar{x} = 4.8$ and the parameter from the null hypothesis is $\mu_0 = 4$. The standard error is $SE = s/\sqrt{n}$. The standardized test statistic is

$$t = \frac{\bar{x} - \mu_0}{s/\sqrt{n}} = \frac{4.8 - 4}{2.3/\sqrt{15}} = 1.35.$$

This is a two-tail test, so the p-value is two times the area above 1.35 in a t-distribution with $df = 14$. We see that the p-value is $2(0.0992) = 0.1984$. This p-value is larger than any reasonable significance level, so we do not find enough evidence to support the alternative hypothesis that $\mu \neq 4$.

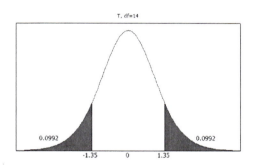

6.141 The null and alternative hypotheses are

$$H_0: \quad \mu = 634$$
$$H_a: \quad \mu \neq 634$$

where μ represents the average number of social ties for a cell phone user. In general, the standardized test statistic is

$$\frac{\text{Sample Statistic} - \text{Null Parameter}}{SE}.$$

In this test for a mean, the sample statistic is $\bar{x} = 664$ and the parameter from the null hypothesis is $\mu_0 = 634$. The standard error is $SE = s/\sqrt{n}$. The standardized test statistic is

$$t = \frac{\bar{x} - \mu_0}{s/\sqrt{n}} = \frac{664 - 634}{778/\sqrt{1700}} = 1.59.$$

This is a two-tail test, so the p-value is two times the area above 1.59 in a t-distribution with df = 1699. (Since n is very large, we could just as easily use the normal distribution to estimate the p-value.) We see that the p-value is $2(0.056) = 0.112$. This p-value is larger than even a 10% significance level, so we do not reject the null hypothesis. There is not sufficient evidence to conclude that US adults who are cell phone users have a different mean number of social ties than the average US adult.

6.143 The sample size of $n = 32$ is large enough to justify using the t-distribution. The null and alternative hypotheses are

$$H_0: \quad \mu = 0$$
$$H_a: \quad \mu > 0$$

where μ represents the mean difference in number of pigeons. In general, the standardized test statistic is

$$\frac{\text{Sample Statistic} - \text{Null Parameter}}{SE}.$$

In this test for a mean, the sample statistic is $\bar{x} = 3.9$ and the parameter from the null hypothesis is $\mu_0 = 0$. The standard error is $SE = s/\sqrt{n}$. The standardized test statistic is

$$t = \frac{\bar{x} - \mu_0}{s/\sqrt{n}} = \frac{3.9 - 0}{6.8/\sqrt{32}} = 3.24.$$

This is an upper-tail test, so the p-value is the area above 3.24 in a t-distribution with df $= 31$. Using technology, we see that the p-value is 0.0014. This p-value is very small so we reject the null hypothesis. There is strong evidence that the mean number of pigeons is higher for the neutral person, even after they switch clothes and behave the same. This suggests that pigeons can recognize faces and hold a grudge.

6.145   (a) The cookies were bought from locations all over the country to try to avoid sampling bias.

(b) Let μ be the mean number of chips per bag. We are testing $H_0 : \mu = 1000$ vs $H_a : \mu > 1000$. The test statistic is
$$t = \frac{1261.6 - 1000}{117.6/\sqrt{42}} = 14.4$$

We use a t-distribution with 41 degrees of freedom. The area to the left of 14.4 is negligible, and p-value ≈ 0. We conclude, with very strong evidence, that the average number of chips per bag of Chips Ahoy! cookies is greater than 1000.

(c) No! The test in part (b) gives convincing evidence that the *average* number of chips per bag is greater than 1000. However, this does not necessarily imply that every individual bag has more than 1000 chips.

6.147 In each of parts (a), (b), and (c), the hypotheses are:

$$H_0: \quad \mu = 265$$
$$H_a: \quad \mu > 265$$

where μ stands for the mean cost of a house in the relevant state, in thousands of dollars.

(a) For New York, we calculate the test statistic

$$t = \frac{\bar{x} - \mu_0}{s/\sqrt{n}} = \frac{565.6 - 265}{697.6/\sqrt{30}} = 2.36$$

Using an upper-tail test with a t-distribution with 29 degrees of freedom, we get a p-value of 0.0126. At a 5% level, we reject the null hypothesis. The average cost of a house in New York State is significantly more then the US average.

(b) For New Jersey, we calculate the test statistic

$$t = \frac{\overline{x} - \mu_0}{s/\sqrt{n}} = \frac{388.5 - 265}{224.7/\sqrt{30}} = 3.01$$

Using an upper-tail test with a t-distribution with 29 degrees of freedom, we get a p-value of 0.0027. We reject the null hypothesis and find strong evidence that the average cost of a house in New Jersey is significantly more then the US average.

(c) Note that the sample mean for Pennsylvania, $\overline{x} = 249.6$ is less than the US average, $\mu = 265$. We don't need a statistical test to see that the average cost in Pennsylvania is not significantly greater than the US average. However if we conducted the test we would have

$$t = \frac{\overline{x} - \mu_0}{s/\sqrt{n}} = \frac{249.6 - 265}{179.3/\sqrt{30}} = -0.47$$

Using an upper-tail test with a t-distribution with 29 degrees of freedom, we get a p-value of 0.6791. (Notice that we still use the upper-tail, which gives a p-value in this case that is greater than 0.5.)

(d) New Jersey has the most evidence that the state average is greater then the US average (smallest p-value). This may surprise you, since the mean for NY homes is greater. However, the standard deviation is also important in determining evidence for or against a claim.

6.149 (a) We see that $n = 30$ with $\overline{x} = 0.25727$ and $s = 0.01039$.

(b) The hypotheses are:

$$H_0 : \quad \mu = 0.250$$
$$H_a : \quad \mu \neq 0.250$$

where μ represents the mean of all team batting averages in Major League Baseball. We calculate the test statistic

$$t = \frac{\overline{x} - \mu_0}{s/\sqrt{n}} = \frac{0.25727 - 0.250}{0.01039/\sqrt{30}} = 3.83$$

We use a t-distribution with 29 degrees of freedom to see that the area above 3.83 is 0.0003. Since this is a two-tail test, the p-value is 2(0.0003) = 0.0006. We reject the null hypothesis, and conclude that the average team batting average is different from (and greater than) 0.250.

(c) The test statistic matches the computer output exactly and the p-value is the same up to rounding off.

6.151 (a) The hypotheses are:

$$H_0 : \quad \mu = 1.0$$
$$H_a : \quad \mu < 1.0$$

where μ represents the mean mercury level of fish in all Florida lakes. Some computer output for the test is shown:

```
One-Sample T: Avg_Mercury
Test of mu = 1 vs < 1
                                        95% Upper
Variable      N    Mean   StDev  SE Mean    Bound      T      P
Avg_Mercury  53  0.5272  0.3410   0.0468   0.6056  -10.09  0.000
```

We see that the p-value is approximately 0, so there is strong evidence that the mean mercury level of fish in Florida lakes is less than 1.0 ppm.

(b) The hypotheses are:

$$H_0: \quad \mu = 0.5$$
$$H_a: \quad \mu < 0.5$$

where μ represents the mean mercury level of fish in all Florida lakes. Some computer output for the test is shown:

```
One-Sample T: Avg_Mercury
Test of mu = 0.5 vs < 0.5
                                           95% Upper
Variable        N    Mean    StDev  SE Mean    Bound     T     P
Avg_Mercury    53  0.5272   0.3410   0.0468   0.6056   0.58  0.718
```

We see that the p-value is 0.718, so there is no evidence at all that the mean is less than 0.5. (In fact, we see that the sample mean, $\bar{x} = 0.5272$ ppm, is actually more than 0.5.)

Section 6.7 Solutions

6.153   (a) The differences in sample proportions will have a mean of $p_A - p_B = 0.70 - 0.60 = 0.10$, and a standard error of

$$SE = \sqrt{\frac{p_A(1 - p_A)}{n_A} + \frac{p_B(1 - p_B)}{n_B}} = \sqrt{\frac{0.70(0.30)}{50} + \frac{0.60(0.40)}{75}} = 0.086$$

(b) We check the sample size for Group A: $n_A p_A = 50(0.70) = 35$ and $n_A(1 - p_A) = 50(0.30) = 15$, and for Group B: $n_B p_B = 75(0.60) = 45$ and $n_B(1 - p_B) = 75(0.40) = 30$. In both cases, the sample size is large enough and the normal distribution applies. The distribution of differences of proportions will be N(0.10, 0.086). We use technology to sketch the graph, or we can sketch it by hand, noting that the differences in sample proportions will be centered at 0.10 and roughly 95% of the distribution lies within two standard deviations on either side of the center. This goes between $0.10 \pm 2(0.086)$ or -0.072 to 0.272. Notice that, although a single proportion will never be negative, the difference in two proportions can be negative (and, in this case, will be negative whenever the sample proportion from Group A is less than the sample proportion from Group B.)

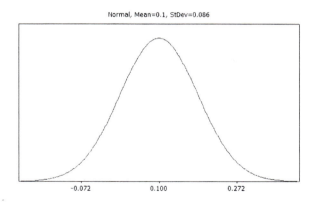

6.155   (a) The differences in sample proportions will have a mean of $p_A - p_B = 0.20 - 0.30 = -0.10$, and a standard error of

$$SE = \sqrt{\frac{p_A(1 - p_A)}{n_A} + \frac{p_B(1 - p_B)}{n_B}} = \sqrt{\frac{0.20(0.80)}{100} + \frac{0.30(0.70)}{50}} = 0.076$$

(b) We check the sample size for Group A: $n_A p_A = 100(0.20) = 20$ and $n_A(1 - p_A) = 100(0.80) = 80$, and for Group B: $n_B p_B = 50(0.30) = 15$ and $n_B(1 - p_B) = 50(0.7) = 35$. The sample sizes are large enough for the normal distribution to apply. The distribution of differences of proportions will be N(-0.10, 0.076). We use technology to sketch the graph, or we can sketch it by hand, noting that the differences in sample proportions will be centered at -0.10 and roughly 95% of the distribution lies within two standard deviations on either side of the center. This goes between $-0.10 \pm 2(0.076)$ or -0.252 to 0.052. Notice that, although a single proportion will never be negative, the difference in two proportions can be negative (and, in this case, will be negative whenever the sample proportion from Group A is less than the sample proportion from Group B.)

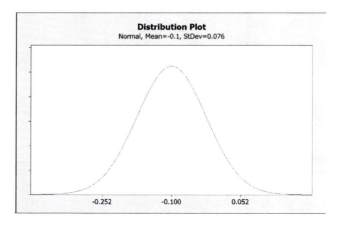

6.157 (a) The differences in sample proportions will have a mean of $p_A - p_B = 0.30 - 0.24 = 0.06$, and a standard error of

$$SE = \sqrt{\frac{p_A(1 - p_A)}{n_A} + \frac{p_B(1 - p_B)}{n_B}} = \sqrt{\frac{0.30(0.70)}{40} + \frac{0.24(0.76)}{30}} = 0.106$$

   (b) We check the sample size for Group A: $n_A p_A = 40(0.30) = 12$ and $n_A(1 - p_A) = 40(0.70) = 28$, and for Group B: $n_B p_B = 30(0.24) = 7.2$. Since $n_B p_B < 10$, the normal distribution does not apply in this case. For inference on the difference in sample proportions in this case, we should use bootstrap or randomization methods.

6.159 (a) This compares two proportions (PC user vs. Mac user) drawn from the same group (students). The methods of this section do not apply to this type of difference in proportions.

   (b) This compares proportions (study abroad) for two different groups (public vs. private). The methods of this section are appropriate for this type of difference in proportions.

   (c) This compares two proportions (in-state vs. out-of-state) drawn from the same group (students). The methods of this section do not apply to this type of difference in proportions.

   (d) This compares proportions (get financial aid) from two different groups (in-state vs. out-of-state). The methods of this section are appropriate for this type of difference in proportions.

6.161 The differences in sample proportions will have a mean of $p_A - p_D = 0.170 - 0.159 = 0.011$, and a standard deviation equal to the standard error SE. We have

$$SE = \sqrt{\frac{p_A(1 - p_A)}{n_A} + \frac{p_D(1 - p_D)}{n_D}} = \sqrt{\frac{0.17(0.83)}{200} + \frac{0.159(0.841)}{200}} = 0.037$$

6.163 The sample size is large enough to use the normal approximation. The mean is

$$p_a - p_{ri} = 0.520 - 0.483 = 0.037$$

the standard error is

$$\sqrt{\frac{0.52(1 - 0.52)}{300} + \frac{0.483(1 - 0.483)}{300}} = 0.041$$

The distribution of $\hat{p}_a - \hat{p}_{ri}$ is approximately $N(0.037, 0.041)$.

6.165 (a) The differences in sample proportions will be centered at the difference in population means, so will have a mean of $p_E - p_J = 0.573 - 0.216 = 0.357$, and a standard deviation equal to the standard error SE. We have

$$SE = \sqrt{\frac{p_E(1 - p_E)}{n_E} + \frac{p_J(1 - p_J)}{n_J}} = \sqrt{\frac{0.573(0.427)}{400} + \frac{0.216(0.784)}{400}} = 0.032$$

(b) The sample sizes of 400 are large enough for the normal distribution to apply. The distribution of differences of proportions will be N(0.357, 0.032). We use technology to sketch the graph, or we can sketch it by hand, noting that the differences in sample proportions will be centered at 0.357 and roughly 95% of the distribution lies within two standard deviations on either side of the center. This goes between $0.357 \pm 2(0.032)$ or 0.293 to 0.421.

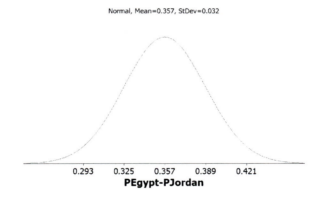

Normal, Mean=0.357, StDev=0.032

0.293   0.325   0.357   0.389   0.421

PEgypt-PJordan

(c) We see from the figure that 0.4 is a plausible sample difference (it's less than 2 SE's above the center) while 0.5 is very unlikely (it's more than 4 SE's above the center).

6.167 Using *StatKey* or other technology to create a bootstrap distribution of the differences in sample proportions, we see for one set of 1000 simulations that $SE = 0.048$. (Answers may vary slightly with other simulations.)

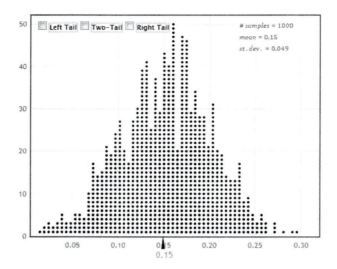

Using the formula with $\hat{p}_A = 90/120 = 0.75$ and $\hat{p}_B = 180/300 = 0.60$ as estimates for the two population proportions, we have

$$SE = \sqrt{\frac{p_A(1 - p_A)}{n_A} + \frac{p_B(1 - p_B)}{n_B}} \approx \sqrt{\frac{0.75(0.25)}{120} + \frac{0.60(0.40)}{300}} = 0.049$$

We see that the bootstrap standard error and the formula match very closely.

6.169 (a) The distribution of $\hat{p}_f - \hat{p}_m$ should be centered at the population difference, $p_f - p_m = 0.255 - 0.214 = 0.041$. The standard error is given by

$$SE = \sqrt{\frac{0.255(1 - 0.255)}{200} + \frac{0.214(1 - 0.214)}{200}} = 0.0423$$

Since $200 \cdot p$ and $200 \cdot (1 - p)$ are both clearly more than 10 for both population proportions we can use a normal distribution to describe these proportions, $\hat{p}_f - \hat{p}_m \sim N(0.041, 0.0423)$. The plot below is a sketch of this normal distribution (along with the area for part (b)).

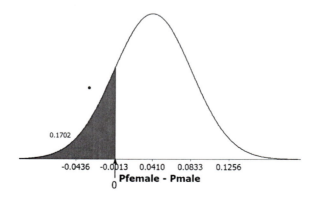

(b) To find the chance that $\hat{p}_f < \hat{p}_m$ we need to find the area below $\hat{p}_f - \hat{p}_m = 0$ in this normal distribution. This is shown on on the figure above, or we can standardize the endpoint of zero to get its z-score

$$z = \frac{0 - 0.041}{0.0423} = -0.97$$

and use the standard normal to find the area below $z = -0.97$ which is 0.166. This means that the sample proportion of college graduates for males will be more than the sample proportion for females in about 17% of all samples of size 200 from both groups, even when the females have a larger proportion in the population.

Section 6.8 Solutions

6.171 The sample sizes are both large enough to use the normal distribution (although just barely large enough in Group 1). For a confidence interval using the normal distribution, we use

$$\text{Sample statistic} \pm z^* \cdot SE.$$

The relevant sample statistic for a confidence interval for a difference in proportions is $\hat{p}_1 - \hat{p}_2 = 0.20 - 0.32$. For a 90% confidence interval, we have $z^* = 1.645$, and we use the sample proportions in computing the standard error. The confidence interval is

$$
\begin{aligned}
(\hat{p}_1 - \hat{p}_2) \quad &\pm \quad z* \cdot \sqrt{\frac{\hat{p}_1(1-\hat{p}_1)}{n_1} + \frac{\hat{p}_2(1-\hat{p}_2)}{n_2}} \\
(0.20 - 0.32) \quad &\pm \quad 1.645 \cdot \sqrt{\frac{0.20(0.80)}{50} + \frac{0.32(0.68)}{100}} \\
-0.12 \quad &\pm \quad 0.121 \\
-0.241 \quad &\text{to} \quad 0.001
\end{aligned}
$$

The best estimate for the difference in the two proportions $p_1 - p_2$ is -0.12, the margin of error is ± 0.121, and the 90% confidence interval for $p_1 - p_2$ is -0.241 to 0.001.

6.173 The sample sizes are both large enough to use the normal distribution. For a confidence interval using the normal distribution, we use

$$\text{Sample statistic} \pm z^* \cdot SE.$$

The relevant sample statistic for a confidence interval for a difference in proportions is $\hat{p}_1 - \hat{p}_2 = 240/500 - 450/1000 = 0.48 - 0.45$. For a 95% confidence interval, we have $z^* = 1.96$, and we use the sample proportions in computing the standard error. The confidence interval is

$$
\begin{aligned}
(\hat{p}_1 - \hat{p}_2) \quad &\pm \quad z* \cdot \sqrt{\frac{\hat{p}_1(1-\hat{p}_1)}{n_1} + \frac{\hat{p}_2(1-\hat{p}_2)}{n_2}} \\
(0.48 - 0.45) \quad &\pm \quad 1.96 \cdot \sqrt{\frac{0.48(0.52)}{500} + \frac{0.45(0.55)}{1000}} \\
0.03 \quad &\pm \quad 0.054 \\
-0.024 \quad &\text{to} \quad 0.084
\end{aligned}
$$

The best estimate for the difference in the two proportions $p_1 - p_2$ is 0.03, the margin of error is ± 0.054, and the 95% confidence interval for $p_1 - p_2$ is -0.024 to 0.084.

6.175 The proportions from the sample in favor of the legislation are

$$\hat{p}_m = 318/520 = 0.612 \text{ for men} \qquad \text{and} \qquad \hat{p}_w = 379/460 = 0.824 \text{ for women}$$

For a 90% confidence interval the normal percentiles are $z = \pm 1.645$. Evaluating the formula for a confidence interval for a difference in proportions gives

$$
\begin{aligned}
(0.612 - 0.824) \quad &\pm \quad 1.645 \cdot \sqrt{\frac{0.612(1-0.612)}{520} + \frac{0.824(1-0.824)}{460}} \\
-0.212 \quad &\pm \quad 1.645 \cdot 0.0278 \\
-0.212 \quad &\pm \quad 0.046 \\
-0.258 \quad &\text{to} \quad -0.166
\end{aligned}
$$

Thus we are 90% sure that the percentage of men who support this gun control legislation is between 25.8% and 16.6% *less* than the percentage of support among women.

6.177 Letting $\hat{p}_e$ and $\hat{p}_w$ represent the proportion of errors in electronic and written prescriptions, respectively, we have

$$\hat{p}_e = \frac{254}{3848} = 0.066 \qquad \text{and} \qquad \hat{p}_w = \frac{1478}{3848} = 0.384$$

The sample sizes are both very large, so it is reasonable to use a normal distribution. For 95% confidence the standard normal endpoint is $z^* = 1.96$. This gives

$$
\begin{aligned}
(\hat{p}_e - \hat{p}_w) \quad &\pm \quad z^* \cdot \sqrt{\frac{\hat{p}_e(1 - \hat{p}_e)}{n_e} + \frac{\hat{p}_w(1 - \hat{p}_w)}{n_w}} \\
(0.066 - 0.384) \quad &\pm \quad 1.96 \cdot \sqrt{\frac{0.066(1 - 0.066)}{3848} + \frac{0.384(1 - 0.384)}{3848}} \\
-0.318 \quad &\pm \quad 0.017 \\
-0.335 \quad &\text{to} \quad -0.301
\end{aligned}
$$

The margin of error is very small because the sample size is so large. Note that if we had subtracted the other way to find a confidence interval for $p_w - p_e$, the interval would be 0.301 to 0.335, with the same interpretation. In each case, we are 95% sure that the error rate is between 0.335 and 0.301 less for electronic prescriptions. This is a very big change! Since zero is not in this interval, it is not plausible that there is no difference. We can be confident that there are fewer errors with electronic prescriptions.

6.179 We have $\hat{p}_M = 0.32$ and $\hat{p}_E = 0.44$. The 95% confidence interval is

$$
\begin{aligned}
(\hat{p}_M - \hat{p}_E) \quad &\pm \quad z^* \cdot \sqrt{\frac{\hat{p}_M(1 - \hat{p}_M)}{n_M} + \frac{\hat{p}_E(1 - \hat{p}_E)}{n_E}} \\
(0.32 - 0.44) \quad &\pm \quad 1.96 \cdot \sqrt{\frac{0.32(0.68)}{122} + \frac{0.44(0.56)}{160}} \\
-0.12 \quad &\pm \quad 0.11 \\
-0.23 \quad &\text{to} \quad -0.01
\end{aligned}
$$

We are 95% sure that breeding success is between 0.23 and 0.01 *less* for metal tagged penguins. This shows a significant difference at a 5% level.

6.181 These are large sample sizes so the normal distribution is appropriate. For 95% confidence $z^* = 1.96$. The sample proportions are $\hat{p}_H = 164/8506 = 0.0193$ and $\hat{p}_P = 122/8102 = 0.0151$ in the HRT group and the placebo group, respectively. The confidence interval is

$$(0.0193 - 0.0151) \pm 1.96 \cdot \sqrt{\frac{0.0193(1 - 0.0193)}{8506} + \frac{0.0151(1 - 0.0151)}{8102}} = 0.0042 \pm 0.0039 = (0.0003, 0.0081).$$

We are 95% sure that the proportion of women who get cardiovascular disease is between 0.0003 and 0.0081 higher among women who get hormone replacement therapy rather than a placebo.

6.183 These are large sample sizes so the normal distribution is appropriate. For 95% confidence $z^* = 1.96$. The sample proportions are $\hat{p}_H = 502/8506 = 0.0590$ and $\hat{p}_P = 458/8102 = 0.0565$ in the HRT group and the placebo group, respectively. The confidence interval is

$$(0.0590 - 0.0565) \pm 1.96 \cdot \sqrt{\frac{0.0590(1 - 0.0590)}{8506} + \frac{0.0565(1 - 0.0565)}{8102}} = 0.0025 \pm 0.0071 = (-0.0046, 0.0096).$$

We are 95% sure that the proportion of women who get any form of cancer is between 0.0046 lower and 0.0096 higher among women who get hormone replacement therapy rather than a placebo.

6.185 We use *StatKey* or other technology to create a bootstrap distribution with at least 1000 simulated differences in proportion. We find the endpoints that contain 95% of the simulated statistics and see that this 95% confidence interval is 0.120 to 0.179.

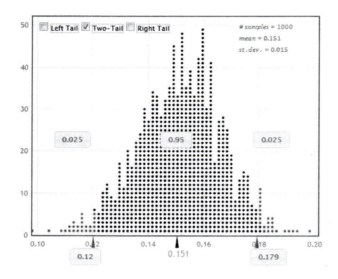

Using the normal distribution and the formula for standard error, we have

$$(0.87 - 0.72) \pm 1.96 \cdot \sqrt{\frac{0.87(0.13)}{800} + \frac{0.72(0.28)}{2252}} = 0.15 \pm 0.030 = (0.12, 0.18)$$

The two methods give very similar confidence intervals.

6.187 We find the proportion who die for those with an infection $\hat{p}_I$ and those without $\hat{p}_N$. Using technology, we have $\hat{p}_I = 0.286$ with $n_I = 84$ and $\hat{p}_N = 0.138$ with $n_N = 116$. Also using technology, we find the 95% confidence interval for $p_I - p_N$ to be 0.033 to 0.263. The proportion who die is between 0.033 and 0.263 higher for those with an infection at admission.

Section 6.9 Solutions

6.189 (a) For Group 1, the proportion who voted is $\hat{p}_1 = 45/70 = 0.643$. For Group 2, the proportion who voted is $\hat{p}_2 = 56/100 = 0.56$. For the pooled proportion, we combine the two groups and look at the proportion who voted. The combined group has $70 + 100 = 170$ people in it, and $45 + 56 = 101$ of them voted, so the pooled proportion is $\hat{p} = 101/170 = 0.594$.

(b) We are testing for a difference in proportions, so we have $H_0 : p_1 = p_2$ vs $H_a : p_1 \neq p_2$. The sample sizes are large enough to use the normal distribution. In general, the standardized test statistic is

$$z = \frac{\text{Sample Statistic} - \text{Null Parameter}}{SE}.$$

In this test for a difference in proportions, the sample statistic is $\hat{p}_1 - \hat{p}_2$ and the parameter from the null hypothesis is 0, since we have $H_0 : p_1 - p_2 = 0$. The standard error uses the pooled proportion. The standardized test statistic is

$$z = \frac{\hat{p}_1 - \hat{p}_2}{\sqrt{\frac{\hat{p}(1-\hat{p})}{n_1} + \frac{\hat{p}(1-\hat{p})}{n_2}}} = \frac{0.643 - 0.56}{\sqrt{\frac{0.594(0.406)}{70} + \frac{0.594(0.406)}{100}}} = 1.08.$$

This is a two-tail test, so the p-value is two times the area above 1.08 in a standard normal distribution. Using technology or a table, we see that the p-value is $2(0.140) = 0.280$. This p-value is quite large so we do not find evidence of any difference between the two groups in the proportion who voted.

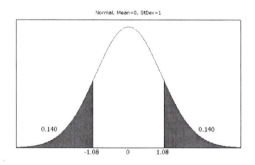

6.191 (a) We'll use subscripts Y for yes these people have the genetic marker and N for no these people do not have the genetic marker. For people with the genetic marker, the proportion who have had depression is $\hat{p}_Y = 0.38$. For people without this specific genetic marker, the proportion who have had depression is $\hat{p}_N = 0.12$. For the pooled proportion, we need to know not just the proportions, but how many people in each group have had depression. We see that $0.38 \cdot 42 = 16$ people with the genetic marker have had depression and $0.12 \cdot 758 = 91$ people without the genetic marker have had depression. We combine the two groups and compute the proportion who have had depression. The combined group has $42 + 758 = 800$ people in it, and $16 + 91 = 107$ of them have had depression. The pooled proportion is $\hat{p} = 107/800 = 0.134$.

(b) This is a test for a difference in proportions, and we have $H_0 : p_Y = p_N$ vs $H_a : p_Y > p_N$. The sample sizes are (just barely for the first group) large enough to use the normal distribution. In general, the standardized test statistic is

$$z = \frac{\text{Sample Statistic} - \text{Null Parameter}}{SE}.$$

In this test for a difference in proportions, the sample statistic is $\hat{p}_Y - \hat{p}_N$ and the parameter from the null hypothesis is 0, since we have $H_0 : p_Y - p_N = 0$. The standard error uses the pooled proportion. The standardized test statistic is

$$z = \frac{\hat{p}_Y - \hat{p}_N}{\sqrt{\frac{\hat{p}(1-\hat{p})}{n_Y} + \frac{\hat{p}(1-\hat{p})}{n_N}}} = \frac{0.38 - 0.12}{\sqrt{\frac{0.134(0.866)}{42} + \frac{0.134(0.866)}{758}}} = 4.82$$

This is an upper-tail test, so the p-value is the area above 4.81 in a standard normal distribution. The test statistic 4.81 is almost five standard deviations above the mean, and the area beyond that is minuscule. The p-value is essentially zero, and we find very strong evidence that people with this specific genetic marker are more likely to suffer from clinical depression.

6.193   (a) For Airline A, the proportion arriving late is $\hat{p}_A = 151/700 = 0.216$. For Airline B, the proportion arriving late is $\hat{p}_B = 87/500 = 0.174$. For the pooled proportion, we combine the two groups and look at the proportion arriving late for the combined group. The combined group has $700 + 500 = 1200$ flights in it, and $151 + 87 = 238$ of them arrived late, so the pooled proportion is $\hat{p} = 238/1200 = 0.198$.

(b) This is a test for a difference in proportions, and we have $H_0 : p_A = p_B$ vs $H_a : p_A \neq p_B$. The sample sizes are very large and we can use the normal distribution. In general, the standardized test statistic is

$$z = \frac{\text{Sample Statistic} - \text{Null Parameter}}{SE}.$$

In this test for a difference in proportions, the sample statistic is $\hat{p}_A - \hat{p}_B$ and the parameter from the null hypothesis is 0, since we have $H_0 : p_A - p_B = 0$. The standard error uses the pooled proportion. The standardized test statistic is

$$z = \frac{\hat{p}_A - \hat{p}_B}{\sqrt{\frac{\hat{p}(1-\hat{p})}{n_A} + \frac{\hat{p}(1-\hat{p})}{n_B}}} = \frac{0.216 - 0.174}{\sqrt{\frac{0.198(0.802)}{700} + \frac{0.198(0.802)}{500}}} = 1.80.$$

This is a two-tail test, so the p-value is two times the area above 1.80 in a standard normal distribution. Using technology or a table, we see that the p-value is $2(0.0359) = 0.072$. The p-value is larger than 0.05, so the results are significant at a 10% level but not at a 5% level. At a 5% level, we do not find evidence of a difference between the airlines in the proportion of flights that are late.

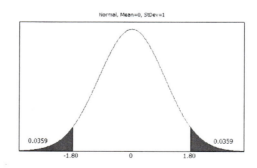

6.195 The hypotheses are

$$\begin{aligned} H_0 : & \quad p_1 = p_2 \\ H_a : & \quad p_1 > p_2 \end{aligned}$$

where p_1 represents the proportion of babies imitating an adult considered reliable and p_2 represents the proportion of babies imitating an adult considered unreliable. There are at least 10 in each group (just barely) so we may use the normal distribution. We compute the sample proportions and the pooled sample proportion:

$$\hat{p}_1 = \frac{18}{30} = 0.60 \qquad \hat{p}_2 = \frac{10}{30} = 0.333 \qquad \hat{p} = \frac{28}{60} = 0.467$$

The standardized test statistic is

$$\frac{\text{Sample statistic} - \text{Null parameter}}{SE} = \frac{(0.60 - 0.333) - 0}{\sqrt{\frac{0.467(0.533)}{30} + \frac{0.467(0.533)}{30}}} = 2.073.$$

This is an upper-tail test, so the p-value is the area above 2.073 in a standard normal distribution. Using technology we see that the p-value is 0.0191. This p-value is less than the 0.05 significance level, so we reject H_0. There is evidence that babies are more likely to imitate an adult who they believe is reliable. (And this effect is seen after one instance of being unreliable. Imagine the impact of repeated instances over time. It pays to be trustworthy!)

6.197 (a) About $800 \times 0.28 = 224$ Quebecers and $500 \times 0.18 = 90$ Texans wanted to separate.

(b) The pooled proportion is

$$\hat{p} = \frac{224 + 90}{800 + 500} \approx 0.242$$

(c) If we let p_1 and p_2 represent the proportion of Quebecers and Texans who want to secede, respectively, the relevant hypotheses are $H_0 : p_1 = p_2$ vs. $H_a : p_1 \neq p_2$. We compute the standardized test statistic as

$$z = \frac{0.28 - 0.18}{\sqrt{\frac{0.242(1-0.242)}{800} + \frac{0.242(1-0.242)}{500}}} = 4.10$$

The p-value is twice the area of the standard normal tail beyond 4.10, which is very small ($p < 0.0001$). Thus, we have very strong evidence that a greater proportion of Quebecers support secession from Canada than Texans support secession from the United States.

6.199 Since the number in the control group that solved the problem is only 4 (which is definitely less than 10) and only 8 in the electrical stimulation group did not solve it, the sample size is too small to use the normal distribution for this test. A randomization test is more appropriate for this data.

6.201 (a) Using I for the internet users and N for the non-internet users, we see that $\hat{p}_I = 807/1754 = 0.46$ and $\hat{p}_N = 130/483 = 0.27$. In the sample, the internet users are more trusting.

(b) A hypothesis test is used to determine whether we can generalize the results from the sample to the population. Since we are looking for a difference, this is a two-tail test. The hypotheses are

$$H_0 : \quad p_I = p_N$$
$$H_a : \quad p_I \neq p_N$$

In addition to the two sample proportions computed in part (a), we compute the pooled proportion. A total of $807 + 130$ people in the full sample of $1754 + 483$ people agreed with the statement, so

$$\text{Pooled proportion} = \hat{p} = \frac{807 + 130}{1754 + 483} = \frac{937}{2237} = 0.419.$$

The standardized test statistic is

$$z = \frac{\hat{p}_I - \hat{p}_N}{\sqrt{\frac{\hat{p}(1-\hat{p})}{n_I} + \frac{\hat{p}(1-\hat{p})}{n_N}}} = \frac{0.46 - 0.27}{\sqrt{\frac{0.419(0.581)}{1754} + \frac{0.419(0.581)}{483}}} = 7.49.$$

This is a two-tail test, so the p-value is two times the area above 7.49 in a standard normal distribution. However, the area above 7.49 in a standard normal is essentially zero (more than seven standard deviations above the mean!). The p-value is essentially zero. The p-value is extremely small, and gives very strong evidence that internet users are more trusting than non-internet users.

(c) No, we cannot conclude that internet uses causes people to be more trusting. The data come from an observational study rather than a randomized experiment. There are many possible confounding factors.

(d) Yes. Level of formal education is a confounding factor if education level affects whether a person uses the internet (it does – more education is associated with more internet use) and also if education level affects how trusting someone is (it does – more education is associated with being more trusting). Remember that a confounding factor is a factor that influences both of the variables of interest. (In fact, even after controlling for education level and several other confounding variables, the data still show that internet users are more trusting than non-users.)

6.203 (a) The sample for New York has only $0.267 \times 30 = 8$ homes with more then 3 bedrooms, so the normal distribution may not be appropriate.

(b) Because the normal distribution may not apply, we return to the methods of Chapter 4 and perform a randomization test for the hypotheses $H_0 : p_{NY} = p_{NJ}$ vs $H_a : p_{NY} \neq p_{NJ}$. We use resampling, taking samples with replacement from the pooled data, because the randomness in this problem comes from sampling, not any kind of allocation. Using *StatKey* or other technology we construct a randomization distribution of many differences in proportions for simulated samples of size 30, taken from a "population" with the same proportion of three bedroom homes as we see in the combined sample.

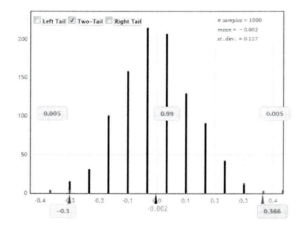

In this randomization distribution we see that 5 of the 1000 samples give differences as extreme as the difference of $0.633 - 0.267 = 0.366$ that occurred in the original sample. Because this is a two-sided test, we double the 0.005 to get p-value = 0.01. We reject the null hypothesis, and conclude that the proportion of homes with more then 3 bedrooms is different between New York and New Jersey.

6.205 These are large sample sizes so the normal distribution is appropriate. The sample proportions are $\hat{p}_H = 164/8506 = 0.0193$ and $\hat{p}_P = 122/8102 = 0.0151$ in the HRT group and the placebo group, respectively. The pooled proportion is

$$\hat{p} = \frac{164 + 122}{8506 + 8102} = 0.0172.$$

The test statistic is

$$z = \frac{0.0193 - 0.0151}{\sqrt{\frac{0.0172(1-0.0172)}{8506} + \frac{0.0172(1-0.0172)}{8102}}} = \frac{0.0042}{0.002} = 2.09$$

Answers may vary slightly due to roundoff. The area in the upper tail of the standard normal distribution is 0.018, which we double to find the p-value of 0.036. Because this was a randomized experiment, we can conclude causality. There is evidence that HRT significantly increases risk of cardiovascular disease.

6.207 These are large sample sizes so the normal distribution is appropriate. The sample proportions are $\hat{p}_H = 502/8506 = 0.0590$ and $\hat{p}_P = 458/8102 = 0.0565$ in the HRT group and the placebo group, respectively. The pooled proportion is

$$\hat{p} = \frac{502 + 458}{8506 + 8102} = 0.0578.$$

The test statistic is

$$z = \frac{0.0590 - 0.0565}{\sqrt{\frac{0.0578(1-0.0578)}{8506} + \frac{0.0578(1-0.0578)}{8102}}} = \frac{0.0025}{0.0036} = 0.69.$$

The area in the upper tail of the standard normal distribution is 0.245, which we double to find the p-value of 0.490. There is not evidence that HRT influences the chance of getting cancer in general.

6.209 The hypotheses are $H_0 : p_m = p_f$ vs $H_a : p_m \neq p_f$ where are p_m and p_f are the proportion with an infection of males and females, respectively. Using technology, we see $\hat{p}_m = 0.411$ with $n_m = 124$ and $\hat{p}_f = 0.434$ with $n_f = 76$. Also using technology, we find that the standardized z-statistic is $z = -0.32$ and the p-value for this two-tailed test is 0.750. There is not sufficient evidence to find a difference between males and females in infection rate.

Section 6.10 Solutions

6.211   (a) The differences in sample means will have a mean of $\mu_1 - \mu_2 = 87 - 81 = 6$ and a standard error of

$$SE = \sqrt{\frac{\sigma_1^2}{n_1} + \frac{\sigma_2^2}{n_2}} = \sqrt{\frac{12^2}{100} + \frac{15^2}{80}} = 2.06$$

  (b) Since $n_1 \geq 30$ and $n_2 \geq 30$, the normal distribution applies and the distribution of differences in sample means will be N(6, 2.06). We use technology to sketch the graph, or we can sketch it by hand, noting that the differences in sample means will be centered at the difference in the population means of 6 and roughly 95% of the distribution lies within two standard deviations on either side of the center. This goes between $6 \pm 2(2.06)$ or 1.88 to 10.12.

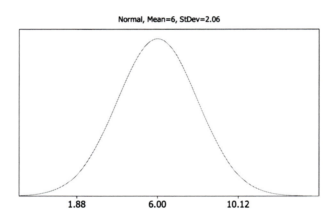

Normal, Mean=6, StDev=2.06

1.88   6.00   10.12

6.213   (a) The differences in sample means will have a mean of $\mu_1 - \mu_2 = 3.2 - 2.8 = 0.4$ and a standard error of

$$SE = \sqrt{\frac{\sigma_1^2}{n_1} + \frac{\sigma_2^2}{n_2}} = \sqrt{\frac{1.7^2}{50} + \frac{1.3^2}{50}} = 0.303.$$

  (b) Since $n_1 \geq 30$ and $n_2 \geq 30$, the normal distribution applies and the distribution of differences in sample means will be N(0.4, 0.30). We use technology to sketch the graph, or we can sketch it by hand, noting that the differences in sample means will be centered at the difference in the population means of 0.4 and roughly 95% of the distribution lies within two standard deviations on either side of the center. This goes between $0.4 \pm 2(0.30)$ or -0.2 to 1.0.

Normal, Mean=0.4, StDev=0.30

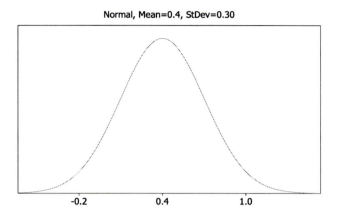

6.215 We use the smaller sample size and subtract 1 to find the degrees of freedom, so we use a t-distribution with df = 14. We see that the values with 2.5% beyond them in each tail are ±2.14.

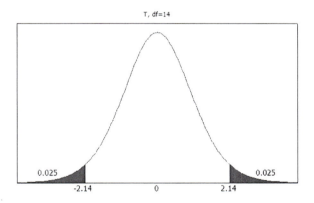

6.217 We use the smaller sample size and subtract 1 to find the degrees of freedom, so we use a t-distribution with df = 29. We see that the probability the t-statistic is less than -1.4 is 0.0861. (With a paper table, we may only be able to specify that the area is between 0.05 and 0.10.)

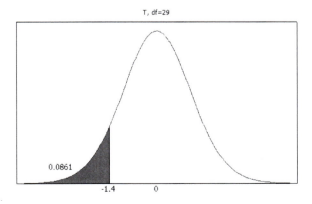

6.219   (a) The differences in sample means will be centered at the difference in population means, so will have a mean of $\mu_m - \mu_f = 534 - 500 = 34$. The standard deviation of the differences in sample means is the standard error, which is

$$SE = \sqrt{\frac{\sigma_m^2}{n_m} + \frac{\sigma_f^2}{n_f}} = \sqrt{\frac{118^2}{40} + \frac{112^2}{60}} = 23.6.$$

(b) The differences in sample means will be centered at the difference in population means, so will have a mean of $\mu_m - \mu_f = 534 - 500 = 34$. The standard deviation of the differences in sample means is the standard error, which is

$$SE = \sqrt{\frac{\sigma_m^2}{n_m} + \frac{\sigma_f^2}{n_f}} = \sqrt{\frac{118^2}{400} + \frac{112^2}{600}} = 7.46.$$

(c) The center stays the same but the spread changes. As the sample sizes go up, the standard error goes down.

6.221 The quantity we are estimating is $\mu_m - \mu_f$, where μ_m represents the average score on this test by all males who took the test and μ_f represents the average score on the test by all females who took the test. The distribution of differences in sample means, $\overline{x}_m - \overline{x}_f$, is centered at 5 which means that the difference in the population means is $\mu_m - \mu_f = 5$. Since $\mu_m - \mu_f$ is positive, we must have $\mu_m > \mu_f$ and find that the population of males scored higher (on average) than females on this test

6.223 The quantity we are estimating is $\mu_N - \mu_E$, where μ_N represents the average score on this test by all non-native English speakers who took the test and μ_E represents the average score on the test by all native English speakers who took the test. The distribution of differences in sample means, $\overline{x}_N - \overline{x}_E$, is centered at 10 which means that the difference in the population means is $\mu_N - \mu_E = 10$. Since $\mu_N - \mu_E$ is positive, we must have $\mu_N > \mu_E$ and find that non-native English speakers scored higher, on average, on the Mathematics test. (This is quite impressive, since the test is given in English!)

6.225 We use *StatKey* or other technology to create a bootstrap distribution. We see for one set of 1000 simulations that $SE \approx 0.77$. (Answers may vary slightly with other simulations.)

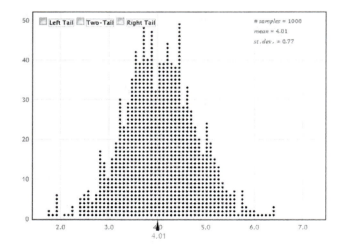

Using the formula from the Central Limit Theorem, and using the sample standard deviations as estimates of the population standard deviations, we have

$$SE = \sqrt{\frac{s_1^2}{n_1} + \frac{s_2^2}{n_2}} = \sqrt{\frac{13.80^2}{500} + \frac{10.75^2}{500}} = 0.782.$$

We see that the bootstrap standard error and the formula match very closely.

6.227 It would be harder for sample means to differ by 3 or more for larger samples, if the population means are the same. The distribution of $\overline{x}_A - \overline{x}_B$ is still centered at zero, and the large samples should give estimates closer to zero, not farther away.

To quantify this we recalculate the standard error to account for the larger samples sizes.

$$SE = \sqrt{\frac{10^2}{100} + \frac{10^2}{100}} = 1.41$$

To find the area beyond ± 3 for a $N(0, 1.41)$ distribution we can use technology such as the figure below, or standardize the endpoints to get $z = \dfrac{\pm 3 - 0}{1.41} = \pm 2.13$.

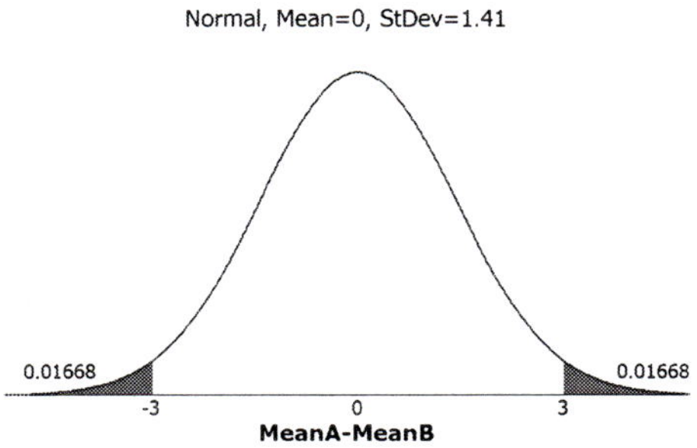

Normal, Mean=0, StDev=1.41

0.01668   0.01668
-3   0   3
MeanA-MeanB

Combining the area for both tails, $0.01668 + 0.01668 = 0.03336$, we find that the class means for samples of size 100 in this scenario will differ by 3 or more points for only a little more than 3% of the common exams.

Section 6.11 Solutions

6.229 For a confidence interval for $\mu_1 - \mu_2$ using the t-distribution, we use

$$(\overline{x}_1 - \overline{x}_2) \pm t^* \sqrt{\frac{s_1^2}{n_1} + \frac{s_2^2}{n_2}}$$

The sample sizes are both 50, so we use a t-distribution with df = 49. For a 90% confidence interval, we have $t^* = 1.68$. The confidence interval is

$$
\begin{aligned}
(10.1 - 12.4) \quad &\pm \quad 1.68 \cdot \sqrt{\frac{2.3^2}{50} + \frac{5.7^2}{50}} \\
-2.3 \quad &\pm \quad 1.46 \\
-3.76 \quad &\text{to} \quad -0.84
\end{aligned}
$$

The best estimate for the difference in means $\mu_1 - \mu_2$ is -2.3, the margin of error is ± 1.46, and the 90% confidence interval for $\mu_1 - \mu_2$ is -3.76 to -0.84. We are 90% confident that the mean for population 2 is between 0.84 and 3.76 more than the mean for population 1.

6.231 For a confidence interval for $\mu_1 - \mu_2$ using the t-distribution, we use

$$(\overline{x}_1 - \overline{x}_2) \pm t^* \sqrt{\frac{s_1^2}{n_1} + \frac{s_2^2}{n_2}}$$

The smaller sample size is 8, so we use a t-distribution with df = 7. For a 95% confidence interval, we have $t^* = 2.36$. The confidence interval is

$$
\begin{aligned}
(5.2 - 4.9) \quad &\pm \quad 2.36 \cdot \sqrt{\frac{2.7^2}{10} + \frac{2.8^2}{8}} \\
0.3 \quad &\pm \quad 3.09 \\
-2.79 \quad &\text{to} \quad 3.39
\end{aligned}
$$

The best estimate for the difference in means $\mu_1 - \mu_2$ is 0.3, the margin of error is ± 3.09, and the 95% confidence interval for $\mu_1 - \mu_2$ is -2.79 to 3.39. We are 95% confident that the difference in the two population means is between -2.79 and 3.39. Notice that zero is in this confidence interval, so it is certainly possible that the two population means are equal.

6.233 For the difference in mean home price between California and New York, we use

$$(\overline{x}_{ca} - \overline{x}_{ny}) \pm t^* \sqrt{\frac{s_{ca}^2}{n_{ca}} + \frac{s_{ny}^2}{n_{ny}}}$$

Both samples have size $n_{ca} = n_{ny} = 30$ so we use 29 degrees of freedom to find $t^* = 1.699$ for 90% confidence. The confidence interval is

$$
\begin{aligned}
(715.1 - 565.6) \quad &\pm \quad 1.699 \sqrt{\frac{1112.3^2}{30} + \frac{697.6^2}{30}} \\
149.5 \quad &\pm \quad 407.3 \\
-257.8 \quad &\text{to} \quad 556.8
\end{aligned}
$$

We are 90% sure that the mean home price in California is somewhere between $257.8 thousand dollars less than in New York to as much as $556.8 thousand more.

6.235 For the difference in mean home price between New York and New Jersey, we use

$$(\overline{x}_{ny} - \overline{x}_{nj}) \pm t^* \sqrt{\frac{s_{ny}^2}{n_{ny}} + \frac{s_{nj}^2}{n_{nj}}}$$

Both samples have size $n_{ny} = n_{nj} = 30$ so we use 29 degrees of freedom to find $t^* = 2.045$ for 95% confidence. The confidence interval is

$$
\begin{aligned}
(565.6 - 388.5) \quad &\pm \quad 2.045 \sqrt{\frac{697.5^2}{30} + \frac{224.7^2}{30}} \\
177.1 \quad &\pm \quad 273.6 \\
-96.5 \quad &\text{to} \quad 450.7
\end{aligned}
$$

We are 95% sure that the mean home price in New York is somewhere between $96.5 thousand dollars less than in New Jersey to as much as $450.7 thousand more.

6.237 We are estimating $\mu_S - \mu_N$ where μ_S represents the mean number of close confidants for those using a social networking site and μ_N represents the mean number for those not using a social networking site. For a 90% confidence interval using a t-distribution with degrees of freedom equal to 946, or a normal distribution since the sample sizes are so large, we use $t^* = 1.65$. We have:

$$
\begin{aligned}
\text{Sample statistic} \quad &\pm \quad t^* \cdot SE \\
(\overline{x}_S - \overline{x}_N) \quad &\pm \quad 1.65 \sqrt{\frac{s_S^2}{n_S} + \frac{s_N^2}{n_N}} \\
(2.5 - 1.9) \quad &\pm \quad 1.65 \sqrt{\frac{1.4^2}{947} + \frac{1.3^2}{1059}} \\
0.6 \quad &\pm \quad 0.10 \\
0.50 \quad &\text{to} \quad 0.70
\end{aligned}
$$

We are 90% confident that the average number of close confidants for a person with a profile on a social networking site is between 0.5 and 0.7 larger than the average number for those without such a profile.

6.239 We estimate the difference in population means using the difference in sample means $\overline{x}_L - \overline{x}_D$, where $\overline{x}_L$ represents the mean weight gain of the mice in light and $\overline{x}_D$ represents the mean weight gain of mice in darkness. For a 99% confidence interval with degrees of freedom equal to 7, we use $t^* = 3.50$. We have:

$$
\begin{aligned}
\text{Sample statistic} \quad &\pm \quad t^* \cdot SE \\
(\overline{x}_L - \overline{x}_D) \quad &\pm \quad t^* \sqrt{\frac{s_L^2}{n_L} + \frac{s_D^2}{n_D}} \\
(9.4 - 5.9) \quad &\pm \quad 3.50 \sqrt{\frac{3.2^2}{19} + \frac{1.9^2}{8}} \\
3.5 \quad &\pm \quad 3.48 \\
0.02 \quad &\text{to} \quad 6.98
\end{aligned}
$$

We are 99% confident that mice with light at night will gain, on average, between 0.02 grams and 6.98 grams more than mice in darkness.

6.241 We let μ_N and μ_Y represent mean GPA for students whose roommate does not bring a videogame (No) or does bring one (Yes) to campus, respectively. The two relevant sample sizes are 88 and 38, so the smallest sample size in these two groups is 38 and the degrees of freedom are $df = 37$. For a 95% confidence level with $df = 37$, we have $t^* = 2.03$. Calculating the 95% confidence interval for the difference in means:

$$
\begin{aligned}
\text{Sample statistic} \quad &\pm \quad t^* \cdot SE \\
(\overline{x}_N - \overline{x}_Y) \quad &\pm \quad t^* \sqrt{\frac{s_N^2}{n_N} + \frac{s_Y^2}{n_Y}} \\
(3.128 - 2.932) \quad &\pm \quad 2.03\sqrt{\frac{0.590^2}{88} + \frac{0.699^2}{38}} \\
0.196 \quad &\pm \quad 0.263 \\
-0.067 \quad &\text{to} \quad 0.459
\end{aligned}
$$

We are 95% sure that, for students who do not bring a videogame to campus, mean GPA for students whose roommate does not bring a videogame will be between 0.067 lower and 0.459 higher than the mean for students whose roommate does bring a videogame. Since 0 is in this interval, it is possible that there is no effect from the roommate bringing a videogame.

6.243 We let μ_N and μ_Y represent mean GPA for students who do not bring a videogame (No) or do bring one (Yes) to campus, respectively. The two relevant sample sizes are 88 and 44, so the smallest sample size in these two groups is 44 and the degrees of freedom are $df = 43$. For a 95% confidence level with $df = 43$, we have $t^* = 2.02$. Calculating the 95% confidence interval for the difference in means:

$$
\begin{aligned}
\text{Sample statistic} \quad &\pm \quad t^* \cdot SE \\
(\overline{x}_N - \overline{x}_Y) \quad &\pm \quad t^* \sqrt{\frac{s_N^2}{n_N} + \frac{s_Y^2}{n_Y}} \\
(3.128 - 3.039) \quad &\pm \quad 2.02\sqrt{\frac{0.590^2}{88} + \frac{0.689^2}{44}} \\
0.089 \quad &\pm \quad 0.245 \\
-0.156 \quad &\text{to} \quad 0.334
\end{aligned}
$$

We are 95% sure that, for students whose roommate does not bring a videogame to campus, mean GPA for students who do not bring a videogame will be between 0.156 lower and 0.245 higher than the mean for students who bring a videogame. Since 0 is in this interval, it is possible that there is no effect from bringing a videogame.

6.245 (a) We let μ_N represent mean GPA for students for whom neither the student nor the roommate brings a videogame and μ_Y represent mean GPA for students for whom both the student and the roommate brings a videogame. The two relevant sample sizes are 88 and 40, so the smallest sample size in these two groups is 40 and the degrees of freedom are $df = 39$. For a 95% confidence level with

$df = 39$, we have $t^* = 2.02$. Calculating the 95% confidence interval for the difference in means:

$$\text{Sample statistic} \quad \pm \quad t^* \cdot SE$$

$$(\overline{x}_N - \overline{x}_Y) \quad \pm \quad t^* \sqrt{\frac{s_N^2}{n_N} + \frac{s_Y^2}{n_Y}}$$

$$(3.128 - 2.754) \quad \pm \quad 2.02 \sqrt{\frac{0.590^2}{88} + \frac{0.639^2}{40}}$$

$$0.374 \quad \pm \quad 0.240$$

$$0.134 \quad \text{to} \quad 0.614$$

We are 95% sure that mean GPA is between 0.134 and 0.240 points higher for students if neither student in the room brings a videogame to campus than if both students in the room bring a videogame to campus. Since 0 is not in this interval and the entire interval is positive, we have fairly convincing evidence that mean GPA is higher if there are no videogames.

(b) We can not conclude that bringing videogames to campus reduces GPA, since these data come from an observational study rather than an experiment. There are many possible confounding variables, such as the possibility that there is a difference in the people who decide to bring a videogame to campus vs those who decide not to.

6.247 (a) The males watch more TV, with a mean in this sample of 7.620 compared to the female mean of 5.237. The difference is $7.620 - 5.237 = 2.383$ hours per week more TV for the males, on average.

(b) We are estimating $\mu_f - \mu_m$, where μ_f represents the number of hours spent watching TV per week by all female students at this university and μ_m represents the number of hours spent watching TV a week for all male students at this university. For a 99% confidence interval with degrees of freedom 168, we use $t^* = 2.61$. The confidence interval is

$$\text{Sample statistic} \quad \pm \quad t^* \cdot SE$$

$$(\overline{x}_f - \overline{x}_m) \quad \pm \quad 1.97 \sqrt{\frac{s_f^2}{n_f} + \frac{s_m^2}{n_m}}$$

$$(5.237 - 7.620) \quad \pm \quad 2.61 \sqrt{\frac{4.100^2}{169} + \frac{6.427^2}{192}}$$

$$-2.383 \quad \pm \quad 1.464$$

$$-3.847 \quad \text{to} \quad -0.919$$

A 99% confidence interval for the difference in means is -3.85 to -0.92.

(c) The confidence intervals are very similar, with minor differences due to round-off error and degrees of freedom. Note you might have switched the order to estimate $\mu_m - \mu_f$ but that would only change the signs of the interval.

(d) We are 99% confident that the average amount males watch TV, in hours per week, is between 3.85 and 0.92 hours more than the amount females watch TV per week, for students at this university.

6.249 Using any statistics package, we see that a 95% confidence interval for $\mu_0 - \mu_1$ is 2.9 to 30.7. We are 95% sure that the difference in average systolic blood pressure between those who live and those who die in the ICU is between 2.9 and 30.7. "No difference" is not plausible, given this sample, since the confidence interval contains only positive differences. For all of the plausible differences, the mean blood pressure is higher for those who lived.

Section 6.12 Solutions

6.251 In general, the standardized test statistic is

$$\frac{\text{Sample Statistic} - \text{Null Parameter}}{SE}$$

In this test for a difference in means, the sample statistic is $\overline{x}_1 - \overline{x}_2$ and the parameter from the null hypothesis is 0 since the null hypothesis statement is equivalent to $\mu_1 - \mu_2 = 0$. Substituting the formula for the standard error for a difference in two means, we compute the t-test statistic to be:

$$t = \frac{(\overline{x}_1 - \overline{x}_2) - 0}{\sqrt{\frac{s_1^2}{n_1} + \frac{s_2^2}{n_2}}} = \frac{15.3 - 18.4}{\sqrt{\frac{11.6^2}{100} + \frac{14.3^2}{80}}} = -1.57$$

This is a two-tailed test, so the p-value is the two times the area below -1.57 in a t-distribution with df $= 79$. We see that the p-value is about $2 \cdot 0.06 = 0.12$. This p-value is not even significant at a 10% level, so we do not find enough evidence to support the alternative hypothesis. There is insufficent evidence that the population means are different.

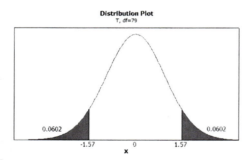

6.253 In general, the standardized test statistic is

$$\frac{\text{Sample Statistic} - \text{Null Parameter}}{SE}$$

In this test for a difference in means, the sample statistic is $\overline{x}_T - \overline{x}_C$ and the parameter from the null hypothesis is 0 since the null hypothesis statement is equivalent to $\mu_T - \mu_C = 0$. Substituting the formula for the standard error for a difference in two means, we compute the t-test statistic to be:

$$t = \frac{(\overline{x}_T - \overline{x}_C) - 0}{\sqrt{\frac{s_T^2}{n_T} + \frac{s_C^2}{n_C}}} = \frac{8.6 - 11.2}{\sqrt{\frac{4.1^2}{25} + \frac{3.4^2}{25}}} = -2.44$$

This is a lower-tail test, so the p-value is the area below -2.44 in a t-distribution with df $= 24$. We see that the p-value is about 0.0112. This p-value shows significance at a 5% level but not quite significant at a 1% level. At a 5% level, we reject H_0 and conclude that there is evidence that $\mu_T < \mu_C$.

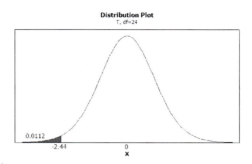

6.255 Let μ_C and μ_I represent mean weight loss after six months for women on a continuous or intermittent calorie restricted diet, respectively. The hypotheses are: The hypotheses are:

$$H_0: \quad \mu_C = \mu_I$$
$$H_a: \quad \mu_C \neq \mu_I$$

The relevant statistic for this test is $\overline{x}_C - \overline{x}_I$, and the relevant null parameter is zero, since from the null hypothesis we have $\mu_C - \mu_I = 0$. The t-test statistic is:

$$t = \frac{\text{Sample statistic} - \text{Null parameter}}{SE} = \frac{(\overline{x}_C - \overline{x}_I) - 0}{\sqrt{\frac{s_C^2}{n_C} + \frac{s_I^2}{n_I}}} = \frac{14.1 - 12.2}{\sqrt{\frac{13.2^2}{54} + \frac{10.6^2}{53}}} = 0.82$$

This is a two-tail test, so the p-value is two times the area above 0.82 in a t-distribution with df = 52. We see that the p-value is $2(0.208) = 0.416$. This p-value is not small at all so we do not reject H_0. We do not see convincing evidence of a difference in effectiveness of the two weight loss methods.

6.257 We first compute the summary statistics

Enriched environment:	$\overline{x}_{EE} = 231.7$	$s_{EE} = 71.2$	$n_{EE} = 7$
Standard environment:	$\overline{x}_{SE} = 438.7$	$s_{SE} = 37.7$	$n_{SE} = 7$

This is a hypothesis test for a difference in means, with μ_{EE} representing the mean number of seconds spent in darkness by mice who have lived in an enriched environment and then been exposed to stress-inducing events, while μ_{SE} represents the same quantity for mice who lived in a standard environment. The hypotheses are:

$$H_0: \quad \mu_{EE} = \mu_{SE}$$
$$H_a: \quad \mu_{EE} < \mu_{SE}$$

The relevant statistic for this test is $\overline{x}_{EE} - \overline{x}_{SE}$, and the relevant null parameter is zero, since from the null hypothesis we have $\mu_{EE} - \mu_{SE} = 0$. The t-test statistic is:

$$t = \frac{\text{Sample statistic} - \text{Null parameter}}{SE} = \frac{(\overline{x}_{EE} - \overline{x}_{SE}) - 0}{\sqrt{\frac{s_{EE}^2}{n_{EE}} + \frac{s_{SE}^2}{n_{SE}}}} = \frac{231.7 - 438.7}{\sqrt{\frac{71.2^2}{7} + \frac{37.7^2}{7}}} = -6.80.$$

This is a lower-tail test, so the p-value is the area to the left of -6.80 in a t-distribution with df = 6. We see that the p-value is 0.00025. This is an extremely small p-value, so we reject H_0, and conclude that there is very strong evidence that mice who have been able to exercise in an enriched environment are better able to handle stress.

6.259 The hypotheses are $H_0 : \mu_C = \mu_W$ vs $H_a : \mu_C > \mu_W$, where μ_C and μ_W are the mean calcium loss after drinking diet cola and water, respectively. The sample sizes are quite small, so we check for extreme skewness or extreme outliers. We see in the dotplots that the data are not too extremely skewed and don't seem to have any extreme outliers, so a t-distribution is acceptable. The t-test statistic is:

$$t = \frac{\text{Sample statistic} - \text{Null parameter}}{SE} = \frac{(\overline{x}_C - \overline{x}_W) - 0}{\sqrt{\frac{s_C^2}{n_C} + \frac{s_W^2}{n_W}}} = \frac{56.0 - 49.1}{\sqrt{\frac{4.93^2}{8} + \frac{3.64^2}{8}}} = 3.18.$$

This is an upper-tail test, so the p-value is the area above 3.18 in a t-distribution with df $= 7$. We see that the p-value is 0.0078. This is a very small p-value, so we reject H_0, and conclude that there is strong evidence (even with such a small sample size) that diet cola drinkers do lose more calcium, on average, than water drinkers. Another reason to drink more water and less diet cola!

6.261 The hypotheses are $H_0 : \mu_M = \mu_E$ vs $H_a : \mu_M > \mu_E$, where μ_M and μ_E represent the mean length of foraging trips in days for metal tagged and electronic tagged penguins, respectively. The sample sizes are large so it is appropriate to use the t-distribution. The t-test statistic is:

$$t = \frac{\text{Sample statistic} - \text{Null parameter}}{SE} = \frac{(\overline{x}_M - \overline{x}_E) - 0}{\sqrt{\frac{s_M^2}{n_M} + \frac{s_E^2}{n_E}}} = \frac{12.70 - 11.60}{\sqrt{\frac{3.71^2}{344} + \frac{4.53^2}{512}}} = 3.89$$

This is an upper-tail test, so the p-value is the area above 3.89 in a t-distribution with df $= 343$. We see that the p-value is 0.00006, or essentially zero. This is an extremely small p-value, so we reject H_0, and conclude that there is very strong evidence that foraging trips are longer on average for penguins with a metal tag.

6.263 (a) All else being equal, the farther apart the means are, the more evidence there is for a difference in means and the smaller the p-value. Option 2 will give a smaller p-value.

(b) All else being equal, the smaller the standard deviations, the easier it is to detect a difference between two means, so the more evidence there is for a difference in means and the smaller the p-value. Option 2 will give a smaller p-value.

(c) All else being equal, the larger the sample size, the more accurate the results. Since the sample means are different in this case, the larger the sample size, the more evidence there is for a difference in means and the smaller the p-value. Option 1 will give a smaller p-value.

For each of these cases we can also consider the formula for the t-statistic

$$t = \frac{(\overline{x}_1 - \overline{x}_2)}{\sqrt{\frac{s_1^2}{n_1} + \frac{s_2^2}{n_2}}}$$

which is positive for each of the situations. A larger t-statistic will be farther in the tail of the t-distribution and yield a smaller p-value. This occurs when the sample means are farther apart, the standard deviations are smaller, or the sample sizes are larger.

6.265 Let μ_1 be the mean grade for students taking a quiz in the second half of class (Late) and μ_2 be the mean grade for quizzes at the beginning of class (Early). The relevant hypotheses are $H_0 : \mu_1 = \mu_2$ vs $H_a : \mu_1 \neq \mu_2$. The computer output shows a test statistic of $t = 1.87$ and p-value=0.066. This is a somewhat small p-value, but not quite significant at a 5% level. The data from this sample show some evidence that it might be better to have the quiz in the second part of class, but that evidence is not very strong.

6.267 The hypotheses are $H_0 : \mu_f = \mu_m$ vs $H_a : \mu_f \neq \mu_m$, where μ_f and μ_m are the respective mean exercise times in the population. Using any statistics package, we see that the t-statistic is -2.98 and the p-value is 0.003. We reject H_0 and conclude that there is a significant evidence showing a difference in mean number of hours a week spent exercising between males and females.

6.269 To compare mean commute times, the hypothesis are $H_0 : \mu_f = \mu_m$ vs $H_a : \mu_f \neq \mu_m$, where μ_f and μ_m are the mean commute times for all female and male commuters in Atlanta.

Here are summary statistics for the commute times, broken down by sex, for the data in **CommuteAtlanta**.

```
Sex    N   Mean  StDev
F     246  26.8   17.3
M     254  31.3   23.4
```

The value of the t-statistic is $t = \dfrac{26.8 - 31.3}{\sqrt{\frac{17.3^2}{246} + \frac{23.4^2}{254}}} = -2.45$

We find the p-value using a t-distribution with $246 - 1 = 245$ degrees of freedom. The area in the lower tail of this distribution below -2.47 is 0.007. Doubling to account for two tails gives p-value = 0.014. This p-value is smaller than the significance level ($\alpha = 0.05$) so we reject the null hypothesis and conclude that the average commute time for women in Atlanta is different from (and, in fact, less than) the average commute time for men.

Section 6.13 Solutions

6.271 For a confidence interval for the average difference $\mu_d = \mu_1 - \mu_2$ using the t-distribution and paired sample data, we use

$$\overline{x}_d \pm t^* \cdot \frac{s_d}{\sqrt{n_d}}$$

We use a t-distribution with df $= 99$, so for a 90% confidence interval, we have $t^* = 1.66$. The confidence interval is

$$
\begin{array}{ccc}
\overline{x}_d & \pm & t^* \cdot \dfrac{s_d}{\sqrt{n_d}} \\[2mm]
556.9 & \pm & 1.66 \cdot \dfrac{143.6}{\sqrt{100}} \\[2mm]
556.9 & \pm & 23.8 \\[1mm]
533.1 & \text{to} & 580.7
\end{array}
$$

The best estimate for the mean difference $\mu_d = \mu_1 - \mu_2$ is 556.9, the margin of error is 23.8, and the 90% confidence interval for μ_d is 533.1 to 580.7. We are 90% confident that the mean for population or treatment 1 is between 533.1 and 580.7 larger than the mean for population or treatment 2.

6.273 For paired difference in means, we begin by finding the differences $d =$ Situation 1 $-$ Situation 2 for each pair. These are shown in the table below.

Case	Difference
1	-8
2	-3
3	3
4	-16
5	-7
6	10
7	-3
8	-1

The mean of these 8 differences is $\overline{x}_d = -3.13$ with standard deviation $s_d = 7.74$. For a confidence interval for the average difference $\mu_d = \mu_1 - \mu_2$ using the t-distribution and paired sample data, we use

$$\overline{x}_d \pm t^* \cdot \frac{s_d}{\sqrt{n_d}}$$

We use a t-distribution with df $= 7$, so for a 95% confidence interval, we have $t^* = 2.36$. The confidence interval is

$$
\begin{array}{ccc}
\overline{x}_d & \pm & t^* \cdot \dfrac{s_d}{\sqrt{n_d}} \\[2mm]
-3.13 & \pm & 2.36 \cdot \dfrac{7.74}{\sqrt{8}} \\[2mm]
-3.13 & \pm & 6.46 \\[1mm]
-9.59 & \text{to} & 3.33
\end{array}
$$

The best estimate for the mean difference $\mu_d = \mu_1 - \mu_2$ is -3.13, the margin of error is 6.46, and the 95% confidence interval for μ_d is -9.59 to 3.33. We are 95% confident that the difference in means $\mu_1 - \mu_2$ is between -9.59 and $+3.33$. (Notice that 0 is within this confidence interval, so based on this sample data we cannot be sure that there is a difference between the means.)

6.275 In general, the standardized test statistic is

$$\frac{\text{Sample Statistic} - \text{Null Parameter}}{SE}$$

In this test for a paired difference in means, the sample statistic is $\overline{x}_d$ and the parameter from the null hypothesis is 0 since the null hypothesis statement is equivalent to $\mu_1 - \mu_2 = \mu_d = 0$. The standard error is $s_d/\sqrt{n_d}$, and the t-test statistic is:

$$t = \frac{\overline{x}_d - 0}{\frac{s_d}{\sqrt{n_d}}} = \frac{-2.6}{\frac{4.1}{\sqrt{18}}} = -2.69$$

This is a two-tail test so, using a t-distribution with df $= 17$, we multiply the area in the tail below -2.69 by two to obtain a p-value of $2(0.008) = 0.016$. At a 5% level, we reject the null hypothesis and find evidence that the means are not the same.

6.277 For paired difference in means, we begin by finding the differences $d =$ Situation 1 $-$ Situation 2 for each pair. These are shown in the table below.

Difference	5	11	-10	25	-7	0	20	-7	6	7

The mean of these ten differences is $\overline{x}_d = 5.0$ with standard deviation $s_d = 11.6$. In general, the standardized test statistic is

$$\frac{\text{Sample Statistic} - \text{Null Parameter}}{SE}$$

In this test for a paired difference in means, the sample statistic is $\overline{x}_d$ and the parameter from the null hypothesis is 0 since the null hypothesis statement is equivalent to $\mu_1 - \mu_2 = \mu_d = 0$. The standard error is $s_d/\sqrt{n_d}$, and the t-test statistic is:

$$t = \frac{\overline{x}_d - 0}{\frac{s_d}{\sqrt{n_d}}} = \frac{5.0}{\frac{11.6}{\sqrt{10}}} = 1.36$$

This is an upper tail test and the p-value is the area above 1.36, in a t-distribution with df $= 9$. We see that the p-value is 0.103. Even at a 10% level, we do not find a significant difference between the two means. We do not find sufficient evidence to conclude that μ_1 is greater than μ_2.

6.279 This is a matched pairs experiment, since all 50 men get both treatments in random order and we are looking at the differences in the results. We use paired data analysis.

6.281 This is a matched pairs experiment, since the "treatment" students are each matched with a similar "control" student. We use paired data analysis.

6.283 This is a matched pairs experiment since the twins are matched and we are investigating the differences between the twins. We use paired data analysis.

6.285 This is a matched pairs experiment so we work with the differences. We have $\overline{x}_d = 2.4$ with $s_d = 6.3$ and $n_d = 47$. To find a 95% confidence interval (with degrees of freedom 46), we use $t^* = 2.01$. We have

$$
\begin{aligned}
\overline{x}_d \quad &\pm \quad t^* \frac{s_d}{\sqrt{n_d}} \\
2.4 \quad &\pm \quad 2.01 \frac{6.3}{\sqrt{47}} \\
2.4 \quad &\pm \quad 1.85 \\
0.55 \quad &\text{to} \quad 4.25
\end{aligned}
$$

We are 95% confident that the average increase in brain glucose metabolism rate from having a cell phone turned on and pressed to the ear for 50 minutes is between 0.55 and 4.25 μmol/100 g per minute.

6.287 We are testing for a difference in two means from a matched pairs experiment. We can use as our null hypothesis either $H_0 : \mu_{on} = \mu_{off}$ or, equivalently, $H_0 : \mu_d = 0$. The two are equivalent since, using d to represent the differences, we have $\mu_d = \mu_{on} - \mu_{off}$. Using the differences, we have

$$
\begin{aligned}
H_0 : \quad &\mu_d = 0 \\
H_a : \quad &\mu_d > 0
\end{aligned}
$$

Notice that this is a one-tail test since we are specifically testing to see if the metabolism is higher for the "on" condition. The t-test statistic is:

$$
t = \frac{\overline{x}_d - 0}{s_d/\sqrt{n_d}} = \frac{2.4}{6.3/\sqrt{47}} = 2.61
$$

Using the t-distribution with degrees of freedom 46, we find a p-value of 0.006. This provides enough evidence to reject H_0 and conclude that mean brain metabolism is significantly higher when a cell phone is turned on and pressed to the ear.

6.289 To find a 99% confidence interval using this paired difference in means data (with degrees of freedom 49), we use $t^* = 2.68$. We have

$$
\begin{aligned}
\overline{x}_d \quad &\pm \quad t^* \frac{s_d}{\sqrt{n_d}} \\
21.7 \quad &\pm \quad 2.68 \cdot \frac{46.5}{\sqrt{50}} \\
21.7 \quad &\pm \quad 17.6 \\
4.1 \quad &\text{to} \quad 39.3
\end{aligned}
$$

We are 99% confident that the average decrease in testosterone level in a man who has sniffed female tears is between 4.1 and 39.3 pg/ml.

6.291 Because the data are paired, we compute a difference (Quiz pulse − Lecture pulse) for each pair. These differences are displayed in the table and dotplot below.

Student	1	2	3	4	5	6	7	8	9	10	Mean	Std. Dev.
Quiz − Lecture	+2	-1	+5	-8	+1	+20	+15	-4	+9	-12	2.7	9.93

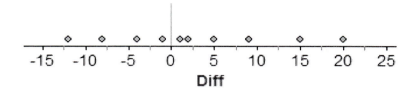

The distribution of differences appears to be relatively symmetric with no clear outliers, so we can use the t-distribution. From the $Quiz - Lecture$ differences in the table above we find that the mean difference is $\overline{x}_d = 2.7$ and the standard deviation of the differences is $s_d = 9.93$. We can express the hypotheses as $H_0 : \mu_Q = \mu_L$ vs $H_a : \mu_Q > \mu_L$, where μ_Q and μ_L are the mean pulse rates during a quiz and lecture, respectively. Equivalently, we can use $H_0 : \mu_d = 0$ vs $H_a : \mu_d > 0$, where μ_d is the mean difference of quiz minus lecture pulse rates. These two ways of expressing the hypotheses are equivalent, since $\mu_d = \mu_Q - \mu_L$, and either way of expressing the hypotheses is acceptable. From the differences and summary statistics, we compute a t-statistic with

$$t = \frac{\overline{x}_d - 0}{s_d/\sqrt{n_d}} = \frac{2.7}{9.93/\sqrt{10}} = 1.52$$

To find the one-tailed p-value for this t-statistic we use the area above 1.52 in a t-distribution with $10 - 1 = 9$ degrees of freedom. This gives a p-value of 0.0814 which is not significant at a 5% level. Thus the data from these 10 students do not provide convincing evidence that mean quiz pulse rate is higher than the mean pulse rate during lecture.

6.293 (a) The data are matched by which story it is, so a matched pairs analysis is appropriate. The matched pairs analysis is particularly important here because there is a great deal of variability between the enjoyment level of the different stories.

(b) We first find the differences for all 12 stories, using the spoiler rating minus the rating for the original.

Story	1	2	3	4	5	6	7	8	9	10	11	12
With spoiler	4.7	5.1	7.9	7.0	7.1	7.2	7.1	7.2	4.8	5.2	4.6	6.7
Original	3.8	4.9	7.4	7.1	6.2	6.1	6.7	7.0	4.3	5.0	4.1	6.1
Difference	0.9	0.2	0.5	-0.1	0.9	1.1	0.4	0.2	0.5	0.2	0.5	0.6

We compute the summary statistics for the differences: $\overline{x}_d = 0.492$ with $s_d = 0.348$ and $n_d = 12$. Using a t-distribution with df = 11 for a 95% confidence level, we have $t^* = 2.20$. We find the 95% confidence interval:

$$\overline{x}_d \quad \pm \quad t^* \cdot \frac{s_d}{\sqrt{n_d}}$$

$$0.492 \quad \pm \quad 2.20 \cdot \frac{0.348}{\sqrt{12}}$$

$$0.492 \quad \pm \quad 0.221$$

$$0.271 \quad \text{to} \quad 0.713$$

We are 95% sure that mean enjoyment rating of a version of a story *with* a spoiler is between 0.271 to 0.713 higher than the mean enjoyment rating of the original story without it.

6.295   (a) This is a difference in means test with separate samples. We first compute the summary statistics and see that, for the first quiz, $n_1 = 10$ with $\bar{x}_1 = 78.6$ with $s_1 = 13.4$ and, for the second quiz, $n_2 = 10$ with $\bar{x} = 84.0$ with $s_2 = 8.1$. To conduct the test for the difference in means, the hypotheses are:

$$H_0 : \quad \mu_1 = \mu_2$$
$$H_a : \quad \mu_1 < \mu_2$$

where μ_1 is the average grade for all students on the first quiz and μ_2 is the average grade for all students on the second quiz. The test statistic is

$$\frac{\bar{x}_1 - \bar{x}_2}{\sqrt{\frac{s_1^2}{n_1} + \frac{s_2^2}{n_2}}} = \frac{78.6 - 84.0}{\sqrt{\frac{13.4^2}{10} + \frac{8.1^2}{10}}} = -1.09$$

Using a t-distribution with df=9, we see that the p-value for this lower tail test is 0.152. This is not significant, even at a 10% level. We do not reject H_0 and do not find convincing evidence that the grades on the second quiz are higher.

(b) This is a paired difference in means test. We begin by finding the differences: first quiz − second quiz.

$$-6 \quad -1 \quad -16 \quad -2 \quad 0 \quad 3 \quad -12 \quad -2 \quad -5 \quad -13$$

The summary statistics for these 10 differences are $n_d = 10$, $\bar{x}_d = -5.4$ with $s_d = 6.29$. The hypotheses are the same as for part (a) and the t-statistic is

$$t = \frac{\bar{x}_d}{s_d/\sqrt{n_d}} = \frac{-5.4}{6.29/\sqrt{10}} = -2.71.$$

Using a t-distribution with df $= 9$, we see that the p-value for this lower-tail test is 0.012. This is significant at a 5% level and almost at even a 1% level. We reject H_0 and find evidence that mean grades on the second quiz are higher.

(c) The spread of the grades is very large on both quizzes, so the high variability makes it hard to find a difference in means with the separate samples. Once we know that the data are paired, it eliminates the variability between people. In this case, it is much better to collect the data using a matched pairs design.

Unit C: Essential Synthesis Solutions

C.1 It is best to begin by adding the totals to the two-way table.

	A	B	C	Total
Yes	21	15	15	51
No	39	50	17	106
Total	60	65	32	157

If we let p be the proportion of bills by Server B, we are testing $H_0 : p = 1/3$ vs $H_a : p > 1/3$. We have $\hat{p} = 65/157 = 0.414$. The sample size is large enough to use the normal distribution. We have

$$z = \frac{\hat{p} - p_0}{\sqrt{\frac{p_0(1-p_0)}{n}}} = \frac{0.414 - 1/3}{\sqrt{\frac{1/3(2/3)}{157}}} = 2.14$$

This is a upper-tail test so the p-value is the area above 2.14 in a standard normal distribution, so we see that the p-value is 0.0162. At a 5% significance level, we reject H_0 and conclude that there is evidence that Server B is responsible for more than 1/3 of the bills at this restaurant. The results, however, are not strong enough to be significant at a 1% level.

C.3 This is a test for a difference in proportions. We are testing $H_0 : p_B = p_C$ vs $H_a : p_B \neq p_C$, where p_B and p_C are the proportions of bills paid with cash for Server B and Server C, respectively. We see from the table that $\hat{p}_B = 50/65 = 0.769$ and $\hat{p} = 17/32 = 0.531$. The sample sizes are large enough to use the normal distribution, and we compute the pooled proportion as $\hat{p} = 67/97 = 0.691$. We have

$$z = \frac{(\hat{p}_B - \hat{p}_C) - 0}{\sqrt{\hat{p}(1 - \hat{p})(\frac{1}{n_B} + \frac{1}{n_C})}} = \frac{0.769 - 0.531}{\sqrt{0.691(0.309)(\frac{1}{65} + \frac{1}{32})}} = 2.39$$

This is a two-tail test, so the p-value is twice the area above 2.39 in a standard normal distribution. We see that

$$\text{p-value} = 2(0.0084) = 0.0168.$$

At a 5% significance level, we reject H_0 and conclude that there is evidence that the proportion paying with cash is not the same between Server B and Server C. Server B appears to have a greater proportion of customers paying with cash. The results, however, are not strong enough to be significant at a 1% level.

C.5 In the sample, the bill is larger when paying with a credit or debit card ($\overline{x}_Y = 29.4 > \overline{x}_N = 19.5$) and there is more variability with a card ($s_Y = 14.5 < s_N = 9.4$). To determine if there is evidence of a difference in the mean bill depending on the method of payment, we do a test for a difference in means. We are testing $H_0 : \mu_Y = \mu_N$ vs $H_a : \mu_Y \neq \mu_N$, where μ_Y represents the mean bill amount when paying with a credit or debit card and μ_N represents the mean bill amount when paying with cash. The sample sizes are large enough to use the t-distribution, even if the underlying bills are not normally distributed. We have

$$t = \frac{(\overline{x}_Y - \overline{x}_N) - 0}{\sqrt{\frac{s_Y^2}{n_Y} + \frac{s_N^2}{n_N}}} = \frac{29.4 - 19.5}{\sqrt{\frac{14.5^2}{51} + \frac{9.4^2}{106}}} = 4.45$$

This is a two-tail test, so the p-value is twice the area above 4.45 in a t-distribution with $51 - 1 = 50$ df. This area is essentially zero, so we have p-value ≈ 0. At any reasonable significance level, we find strong evidence to reject H_0. There is strong evidence that the mean bill amounts are not the same between the two payment methods. Customers with higher bills are more likely to pay with a credit/debit card.

C.7 None of the data appears to be severely skewed or with significant outliers, so it is fine to use the t-distribution. We use technology and the data in **MentalMuscle** to obtain the means and standard deviations for the times in each group as summarized in the table below.

Variable	Action	Fatigue	N	Mean	StDev
Time	Mental	Pre	8	7.34	1.22
Time	Mental	Post	8	6.10	0.95
Time	Actual	Pre	8	7.16	0.70
Time	Actual	Post	8	8.04	1.07

(a) To compare the mean mental and actual times before fatigue we do a two-sample t-test for difference in means. Letting μ_M represent the mean time for someone mentally imaging the actions before any muscle fatigue and μ_A represent the mean time for someone actually performing the actions before any muscle fatigue, our hypotheses are:

$$H_0: \quad \mu_M = \mu_A$$
$$H_a: \quad \mu_M \neq \mu_A$$

Using the pre-fatigue data values in each case, the relevant summary statistics as $\bar{x}_M = 7.34$ with $s_M = 1.22$ and $n_M = 8$ for the mental pre-fatigue group and $\bar{x}_A = 7.16$ with $s_A = 0.70$ and $n_A = 8$ for the actual pre-fatigue group. The test statistic is

$$t = \frac{\bar{x}_M - \bar{x}_A}{\sqrt{\frac{s_M^2}{n_M} + \frac{s_A^2}{n_A}}} = \frac{7.34 - 7.16}{\sqrt{\frac{1.22^2}{8} + \frac{0.70^2}{8}}} = 0.36$$

This is a two-tail test, so the p-value is twice the area above 0.36 in a t-distribution with $df = 7$. We see that the p-value is $2(0.365) = 0.73$. This is a very large p-value, so we do not reject H_0. There is no convincing evidence at all of a difference between the two groups before muscle fatigue.

(b) For each action, the same 8 people perform the movements twice, once before muscle fatigue and once after. To compare the pre-fatigue and post-fatigue means for those actually doing the movements we use a paired difference in means test. We compute the differences, $D = PostFatigue - PreFatigue$, using the data for the people doing the actual movements and see that the 8 differences are:

$$2.5, \quad 0, \quad 0.7, \quad 0.5, \quad 0.2, \quad 0.6, \quad -0.3, \quad 2.8$$

The summary statistics for the differences are $\bar{x}_D = 0.88$ with $s_D = 1.15$ and $n_D = 8$. The hypotheses are

$$H_0: \quad \mu_D = 0$$
$$H_a: \quad \mu_D > 0$$

Notice that if we had subtracted the other direction ($PreFatigue - PostFatigue$), the alternative would be in the other direction and all the differences would have the opposite sign, but the results would be identical. The test statistic is

$$t = \frac{\bar{x}_D}{s_D/\sqrt{n_D}} = \frac{0.88}{1.15/\sqrt{8}} = 2.16$$

This is an upper-tail test, so the p-value is the area above 2.16 in a t-distribution with $df = 7$. Using technology we see that the p-value is 0.0338. At a 5% significance level, we reject H_0 and conclude that people are slower, on average, at performing physical motions when they have muscle fatigue. This is not surprising; we slow down when we are tired!

(c) As in part (b), the same 8 people perform the movements twice, once before muscle fatigue and once after. To compare the pre-fatigue and post-fatigue means for those doing mental movements we use a paired difference in means test. We compute the differences, $D = PostFatigue - PreFatigue$, using the data for the people doing the mental imaging and see that the 8 differences are:

$$1.5, \quad -3.9, \quad -1.5, \quad -1.1, \quad -1.1, \quad -0.2, \quad -1.5, \quad -2.1$$

The summary statistics for the differences are $\overline{x}_D = -1.24$ with $s_D = 1.54$ and $n_D = 8$. The hypotheses are;

$$H_0: \quad \mu_D = 0$$
$$H_a: \quad \mu_D < 0$$

Notice that if we had subtracted the other direction ($PreFatigue - PostFatigue$), the alternative would be in the other direction and all the differences would have the opposite sign, but the results would be identical. The test statistic is

$$t = \frac{\overline{x}_D}{s_D/\sqrt{n_D}} = \frac{-1.24}{1.54/\sqrt{8}} = -2.28$$

This is a lower-tail test, so the p-value is the area below -2.28 in a t-distribution with $df = 7$. Using technology we see that the p-value is 0.0283. At a 5% significance level, we reject H_0 and conclude that people are faster, on average, at mentally imaging physical motions when they have muscle fatigue. This is the new finding from the study: when people have muscle fatigue, they speed up their mental imaging – presumable because they just want to get done! It is likely that this makes the mental imagery less effective.

(d) There are two separate groups (*Mental* vs *Actual*) so we do a difference in means test with two groups. Letting μ_M represent the mean time for someone mentally imaging the actions after muscle fatigue and μ_A represent the mean time for someone actually performing the actions after muscle fatigue, our hypotheses are:

$$H_0: \quad \mu_M = \mu_A$$
$$H_a: \quad \mu_M \neq \mu_A$$

Using the post-fatigue data values in each case, the relevant summary statistics are $\overline{x}_M = 6.10$ with $s_M = 0.95$ and $n_M = 8$ for the mental post-fatigue group and $\overline{x}_A = 8.04$ with $s_A = 1.07$ and $n_A = 8$ for the actual post-fatigue group. The test statistic is

$$t = \frac{\overline{x}_M - \overline{x}_A}{\sqrt{\frac{s_M^2}{n_M} + \frac{s_A^2}{n_A}}} = \frac{6.10 - 8.04}{\sqrt{\frac{0.95^2}{8} + \frac{1.07^2}{8}}} = -3.83.$$

This is a two-tail test, so the p-value is twice the area below -3.83 in a t-distribution with $df = 7$. We see that the p-value is $2(0.0032) = 0.0064$. This is a very small p-value, so we reject H_0. There is strong evidence of a difference in the mean times between the two groups after muscle fatigue.

(e) Before muscle fatigue, the group mentally imaging doing the actions was remarkably similar in time to those actually doing the actions. The mental imaging was quite accurate. However, muscle fatigue caused those actually doing the motions to slow down while it caused those mentally imaging the motions to speed up. Taken together, there was a significant difference between the two groups after experiencing muscle fatigue so that the mental imaging of the motions was not as accurate at matching the actual motions.

C.9 If μ is the mean number of free throws attempted in games by the Miami Heat, we test $H_0 : \mu = 25.0$ vs $H_a : \mu \neq 25.0$. For the sample of 82 games in 2010-11 in **MiamiHeat** we find the mean number of free throw attempts by the Heat is $\bar{x} = 27.9$ with a standard deviation of 7.88. The t-statistic is

$$t = \frac{\bar{x} - \mu_0}{s/\sqrt{n}} = \frac{27.9 - 25.0}{7.88/\sqrt{82}} = 3.33$$

We find the p-value by doubling the area beyond 3.33 in a t-distribution with $82 - 1 = 81$ degrees of freedom, p-value $= 2 \cdot 0.00065 = 0.0013$. (We can also find the p-value using technology.) This is a very small p-value, so we have strong evidence that the Miami Heat get more free throw attempts per game on average than is typical for NBA teams.

C.11 This question is asking for a test to compare two proportions, p_H and p_A, the proportion of free throws the Heat make in home and away games, respectively. The question also suggests a particular direction so the hypotheses are $H_0 : p_H = p_A$ vs $H_a : p_H > p_A$. Based on the sample results we get the proportions below for each location and the combined data.

$$\hat{p}_H = \frac{949}{1222} = 0.7766 \qquad \hat{p}_A = \frac{811}{1066} = 0.7608 \qquad \hat{p} = \frac{1760}{2288} = 0.7692$$

The standardized test statistic is

$$z = \frac{0.7766 - 0.7608}{\sqrt{0.7692(1 - 0.7692)\left(\frac{1}{1222} + \frac{1}{1066}\right)}} = \frac{0.0158}{0.01766} = 0.895$$

Since the sample sizes are large we find a p-value using the area in a N(0,1) distribution that lies beyond $z = 0.895$. This gives p-value$= 0.185$ which is not small, so we do not have sufficient evidence to conclude that the Heat successfully makes a higher proportion of its free throw attempts at home compared to away.

C.13 (a) The hypotheses are

$$H_0 : \quad \mu = 72$$
$$H_a : \quad \mu \neq 72$$

where μ represents the average heart rate of all patients admitted to this ICU. Using technology, we see that the average heart rate for the sample of patients is $\bar{x} = 98.92$ and we see that the p-value for the test is essentially zero (with a t-statistic of 14.2). There is very strong evidence that the average heart rate of ICU patients is not 72.

(b) Using technology, we see that 40 of the 200 patients died, so $\hat{p} = 40/200 = 0.2$. Using technology, we see that a 95% confidence interval for the proportion who die is (0.147, 0.262). We are 95% confident that between 14.7% and 26.2% of the ICU patients die at this hospital.

(c) We see that there were 124 females (62%) and 76 males (38%), so more females were admitted to the ICU in this sample. To test whether genders are equally split, we do a one-sample proportion test. It doesn't matter whether we test for the proportion of males or the proportion of females, since if we know one, we can compute the other, and in either case we are testing whether the proportion is significantly different from 0.5. Using p to denote the proportion of ICU patients that are female, we test

$$H_0 : \quad p = 0.5$$
$$H_a : \quad p \neq 0.5$$

Using technology, we see that the p-value for this test is 0.001. This provides very strong evidence that patients are not evenly split between the genders. There are significantly more females than males admitted to this ICU unit.

(d) This is a difference in means test. If we let μ_M represent the mean age of male ICU patients and μ_F represent the mean age of female ICU patients, the hypotheses are

$$H_0: \quad \mu_M = \mu_F$$
$$H_a: \quad \mu_M \neq \mu_F$$

Using technology, we see that the p-value is 0.184. We do not reject H_0 and do not find convincing evidence that mean age differs between males and females.

(e) This is a difference in proportions test. If we let p_M represent the proportion of males who die and p_F represent the proportion of females who die, the hypotheses are

$$H_0: \quad p_M = p_F$$
$$H_a: \quad p_M \neq p_F$$

Using technology, we see that the p-value is 0.772. We do not reject H_0 and do not find convincing evidence that the proportion who die differs between males and females.

Unit C: Review Exercise Solutions

C.15 Using technology on a calculator or computer, we see that

(a) The area below $z = -2.10$ is 0.018.

(b) The area above $z = 1.25$ is 0.106.

C.17 Using technology on a calculator or computer, we see that the endpoint z is

(a) $z = 0.253$

(b) $z = -2.054$

C.19 We use a t-distribution with df $= 24$. Using technology, we see that the values with 5% beyond them in each tail are ± 1.711.

C.21 We use a t-distribution with df $= 9$. Using technology, we see that the area above 2.75 is 0.011.

C.23   (a) To find a confidence interval using the normal distribution, we use

$$\text{Sample statistic } \pm z^* \cdot SE.$$

We are finding a confidence interval for a mean, so the statistic from the original sample is $\bar{x} = 12.79$. For a 95% confidence interval, we use $z^* = 1.960$ and we have $SE = 0.30$. Putting this information together, we have

$$
\begin{array}{ccc}
\bar{x} & \pm & z^* \cdot SE \\
12.79 & \pm & 1.960 \cdot (0.30) \\
12.79 & \pm & 0.59 \\
12.20 & \text{to} & 13.38
\end{array}
$$

We are 95% confident that the average number of grams of fiber consumed per day is 12.20 grams and 13.38 grams.

(b) The relevant hypotheses are $H_0 : \mu = 12$ vs $H_a : \mu > 12$, where μ is the mean number of grams of fiber consumed per day by all people in the population. The statistic of interest is $\bar{x} = 12.79$. The standard error of this statistic is given as $SE = 0.30$ and the null hypothesis is that the population mean is 12. We compute the standardized test statistic with

$$z = \frac{\text{Sample Statistic} - \text{Null Parameter}}{SE} = \frac{12.79 - 12}{0.30} = 2.63$$

Using technology, the area under a $N(0,1)$ curve beyond $z = 2.63$ is 0.004. This small p-value provides strong evidence that the mean number of grams of fiber consumed per day is greater than 12.

C.25 The z test statistic can be interpreted as a z-score so this test statistic is more than 5 standard deviations out in the tail. This will have very little area beyond it (so the p-value is very small) and will provide strong evidence for the alternative hypothesis (so we reject H_0).

(a) Small

(b) Reject H_0

C.27 The z test statistic can be interpreted as a z-score so this test statistic is less than 1 standard deviation from the mean. This will have quite a large area beyond it (so the p-value is relatively large) and will not provide much evidence for the alternative hypothesis (so we do not reject H_0).

(a) Large

(b) Do not reject H_0

C.29 For relatively large sample size, the t-distribution is very similar to the standard normal distribution, so the t test statistic is similar to a z-score. This test statistic is approximately 7 standard deviations out in the tail! This will have very little area beyond it (so the p-value is very small) and will provide strong evidence for the alternative hypothesis (so we reject H_0).

(a) Small

(b) Reject H_0

C.31 We are estimating mean amount spent in the store, so we use a confidence interval for a mean.

C.33 We are testing for a difference in the proportion classified as insulin-resistant between high sugar and normal diets, so we use a hypothesis test for a difference in proportions.

C.35 We are estimating the difference in the mean financial aid package between two groups of students, so we use a confidence interval for a difference in means.

C.37 We are testing whether the proportion of left-handers is different from 12%, so we use a hypothesis test for a proportion.

C.39 We have $\hat{p} = 0.73$ with $n = 1000$. For a 95% confidence interval, we have $z^* = 1.96$. The 95% confidence interval for a proportion is

$$
\begin{aligned}
\hat{p} \ &\pm\ z^* \cdot \sqrt{\frac{\hat{p}(1-\hat{p})}{n}} \\
0.73 \ &\pm\ 1.96 \cdot \sqrt{\frac{0.73(1-0.73)}{1000}} \\
0.73 \ &\pm\ 0.028 \\
0.702 \ &\text{to}\ 0.758
\end{aligned}
$$

We are 95% confident that the proportion of likely voters in the US who think a woman president is likely in the next 10 years is between 0.702 and 0.758.

C.41 We let μ represent the average scrotal temperature increase for a man with a laptop computer on his lap for an hour. The hypotheses are

$$
\begin{aligned}
H_0 : \ &\mu = 1 \\
H_a : \ &\mu > 1
\end{aligned}
$$

The test statistic is given by

$$
t = \frac{\overline{x} - \mu_0}{s/\sqrt{n}} = \frac{2.31 - 1}{0.96/\sqrt{29}} = 7.35
$$

This is an upper-tail test, so the p-value is the area above 7.35 in a t-distribution with $df = 28$. We see that the p-value is essentially zero. There is very strong evidence that mean scrotal temperature increase is greater than $1°C$.

C.43 We are finding a confidence interval for a difference in proportions. Using p_D for the proportion of people with dyslexia who have the gene disruption and p_C be the poportion of people without dyslexia who have the gene disruption, we want to estimate $p_D - p_C$. Since $\hat{p}_D = 10/109 = 0.092$ and $\hat{p}_C = 5/195 = 0.026$, the sample statistic is $\hat{p}_D - \hat{p}_C = 0.092 - 0.026 = 0.066$. Since there are only 5 in the group of people who have dyslexia and are in the control group, the conditions are not met for using the normal distribution. We use a bootstrap method instead. Using StatKey or other technology, we create a bootstrap distribution of differences in proportions after sampling with replacement from the respective samples.

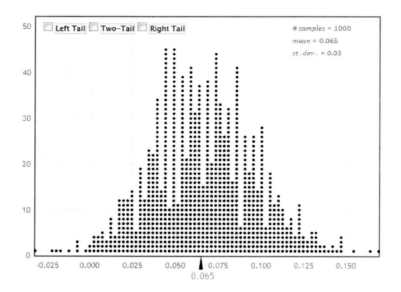

Using the $2 \cdot SE$ method, we estimate the 95% confidence interval to be

$$
\begin{array}{ccc}
\text{Sample statistic} & \pm & 2 \cdot SE \\
0.066 & \pm & 2(0.030) \\
0.066 & \pm & 0.060 \\
0.006 & \text{to} & 0.126
\end{array}
$$

We are 95% confident that the difference in proportions with the gene disruption between people with dyslexia and people without is between 0.006 and 0.126. We could also use percentages from the bootstrap distribution to estimate the confidence interval, and the answer will be similar but possibly not identical.

C.45 This is a hypothesis test for a proportion. There are 96 people total and the lie detector detects lying for $31 + 27 = 58$ of them. We have $\hat{p} = 58/96 = 0.604$. Letting p represent the proportion of times a lie detector says a person is lying regardless of whether or not the person is lying, we have the hypotheses

$$
\begin{aligned}
H_0: & \quad p = 0.5 \\
H_a: & \quad p > 0.5
\end{aligned}
$$

The test statistic is

$$
z = \frac{\hat{p} - p_0}{\sqrt{\frac{p_0(1-p_0)}{n}}} = \frac{0.604 - 0.5}{\sqrt{\frac{0.5(0.5)}{96}}} = 2.04.
$$

This is an upper-tail test so the p-value is the area above 2.04 in a normal distribution. We see that the p-value is 0.0207. At a 5% significance level, we reject H_0 and conclude that a lie detector says a person is lying more than 50% of the time. The results are significant at the 5% level but not at the 1% level.

C.47 This is a confidence interval for a difference in proportions. We need sample proportions (of detected lies) for the lying and truthful groups. We have

$$\hat{p}_L = 31/48 = 0.646 \qquad \text{and} \qquad \hat{p}_T = 27/48 = 0.563$$

For a 95% confidence interval, we have $z^* = 1.96$. The confidence interval is

$$
\begin{aligned}
(\hat{p}_L - \hat{p}_T) \quad &\pm \quad z^* \cdot \sqrt{\frac{\hat{p}_L(1 - \hat{p}_L)}{n_L} + \frac{\hat{p}_T(1 - \hat{p}_T)}{n_T}} \\
(0.646 - 0.563) \quad &\pm \quad 1.96 \cdot \sqrt{\frac{0.646(0.354)}{48} + \frac{0.563(0.437)}{48}} \\
0.083 \quad &\pm \quad 0.195 \\
-0.112 \quad &\text{to} \quad 0.278.
\end{aligned}
$$

We are 95% confident that the lie detector is between -0.112 and 0.278 more likely to detect a person lying if the person actually is lying than if the person is telling the truth.

C.49 (a) For the data from Great Britain, the sample size is only 8 and the data appear to be quite skewed with outliers. A t-distribution is probably not appropriate for these data, so we use a bootstrap method to produce the distribution of sample means shown below.

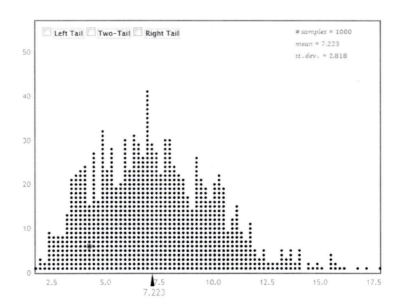

Using the $2 \cdot SE$ method to construct the confidence interval, we see that the standard error for one bootstrap distribution using this data is about 2.82. We have

$$
\begin{aligned}
\overline{x} \ &\pm\ 2 \cdot SE \\
7.21 \ &\pm\ 2(2.82) \\
7.21 \ &\pm\ 5.64 \\
1.57 \ &\text{to}\ 12.85
\end{aligned}
$$

We are 95% confident that the mean arsenic level in toenails of all people living near the former arsenic mine in Great Britain is between 1.57 mg/kg and 12.85 mg/kg.

(b) For the data from New Hampshire the sample size is 19 and the data are only mildly skewed, so we use the t-distribution. (It is also appropriate to use a bootstrap distribution with this data and the results are similar.) For a 95% confidence interval with $df = 18$, we have $t^* = 2.10$. The summary statistics for the data are $\overline{x} = 0.2719$ and $s = 0.2365$. A 95% confidence interval is

$$
\begin{aligned}
\overline{x} \ &\pm\ t^* \cdot \frac{s}{\sqrt{n}} \\
0.2719 \ &\pm\ 2.10 \cdot \frac{0.2365}{\sqrt{19}} \\
0.2719 \ &\pm\ 0.1139 \\
0.1580 \ &\text{to}\ 0.3858
\end{aligned}
$$

We are 95% confident that the mean arsenic level in toenails of people with private wells in New Hampshire is between 0.158 ppm and 0.386 ppm.

C.51 This is a hypothesis test for a difference in means. Using μ_R for the average rating with a red background and μ_W for the average rating with a white background, the hypotheses for the test are:

$$
\begin{aligned}
H_0 : \ & \mu_R = \mu_W \\
H_a : \ & \mu_R > \mu_W
\end{aligned}
$$

The relevant statistic for this test is $\overline{x}_R - \overline{x}_W$, where $\overline{x}_R$ represents the mean rating in the sample with the red background and $\overline{x}_W$ represents the mean rating in the sample with the white background. The relevant null parameter is zero, since from the null hypothesis we have $\mu_R - \mu_W = 0$. The t-test statistic is:

$$
t = \frac{\text{Sample statistic} - \text{Null parameter}}{SE} = \frac{(\overline{x}_R - \overline{x}_W) - 0}{\sqrt{\frac{s_R^2}{n_R} + \frac{s_W^2}{n_W}}} = \frac{7.2 - 6.1}{\sqrt{\frac{0.6^2}{15} + \frac{0.4^2}{12}}} = 5.69
$$

This is a right tail test, so using the t-distribution with df = 11, we see that the p-value is less than 0.0001. This is an extremely small p-value, so we reject H_0, and conclude that there is strong evidence that average attractiveness rating is higher with the red background.

C.53 (a) For the cardiac arrest patients, the proportion is $\hat{p}_A = 11/116 = 0.095$. There are $1595 - 116 = 1479$ other patients and $27 - 11 = 16$ of them reported a near-death experience, so for the patients with other cardiac problems, the proportion is $\hat{p}_B = 16/1479 = 0.011$.

(b) Letting p_A be the proportion of cardiac arrest patients reporting a near-death experience and p_B be the proportion of other heart patients reporting a near-death experience, the hypotheses are

$$
\begin{aligned}
H_0 : \ & p_A = p_B \\
H_a : \ & p_A > p_B
\end{aligned}
$$

The sample sizes are large enough to use the normal distribution. To compute the test statistic, we need three sample proportions. In addition to $\hat{p}_A$ and $\hat{p}_B$ from part (a), we also compute the pooled proportion. We see that $\hat{p} = 27/1595 = 0.017$. The test statistic is

$$z = \frac{\hat{p}_A - \hat{p}_B}{\sqrt{\hat{p}(1-\hat{p})\left(\frac{1}{n_A} + \frac{1}{n_B}\right)}} = \frac{0.095 - 0.011}{\sqrt{0.017(0.983)\left(\frac{1}{116} + \frac{1}{1479}\right)}} = 6.74$$

This is an upper-tail test, so the p-value is the area above 6.74 in a normal distribution. This is almost seven standard deviations above the mean, so we see that the p-value is essentially zero. There is very strong evidence that near-death experiences are significantly more likely for people who have cardiac arrest than for people with other heart problems.

C.55 (a) We combine the *Training* and *Both* categories for those who received training and we combine the *Medication* and *Neither* categories for those who did not receive training. For improvement, we combine the *Much* and *Some* categories for those who received any improvement at all: the *Yes* category. The results are shown in the table.

Any Improvement	Training	No training	Total
Yes	28	10	38
No	9	25	34
Total	37	35	72

(b) This is a test for a difference in proportions. Letting p_T represent the proportion of insomniacs getting any improvement from training and p_N represent the proportion of insomniacs getting any improvement with no training, the hypotheses are

$$H_0: \quad p_T = p_N$$
$$H_a: \quad p_T > p_N$$

To compute the test statistic, we need three sample proportions. We have $\hat{p}_T = 28/37 = 0.757$ and $\hat{p}_N = 10/35 = 0.286$. To compute the pooled proportion, we see that a total of 38 people had some improvement out of 72 people in the study, so the pooled proportion is

$$\hat{p} = \frac{38}{72} = 0.528.$$

The test statistic is

$$\frac{\hat{p}_T - \hat{p}_N}{\sqrt{\hat{p}(1-\hat{p})\left(\frac{1}{n_T} + \frac{1}{n_N}\right)}} = \frac{0.757 - 0.286}{\sqrt{0.528(0.472)\left(\frac{1}{37} + \frac{1}{35}\right)}} = 4.0.$$

This is an upper-tail test, so the p-value is the area above 4.0 in a standard normal distribution. The p-value is 0.00003, or essentially zero. There is very strong evidence that training in behavioral modifications helps older people fight insomnia.

C.57 This is a hypothesis test for a mean, and the data appear to be normal enough that we can use the t-distribution for the test. Letting μ represent the mean body temperature for this person. The hypotheses are

$$H_0: \quad \mu = 98.6$$
$$H_a: \quad \mu \neq 98.6$$

The summary sample statistics are $\bar{x} = 98.4$ with $s = 0.49$ with $n = 12$. The test statistic is

$$\frac{\bar{x} - \mu_0}{\frac{s}{\sqrt{n}}} = \frac{98.4 - 98.6}{\frac{0.49}{\sqrt{12}}} = -1.41.$$

This is a two-tail test so the p-value is twice the area below -1.41 in a t-distribution with $df = 11$. We see that the p-value is $2(0.093) = 0.186$. This is not a small p-value so if these readings are a random sample of the person's body temperatures throughout the day, then there is not convincing evidence that the mean body temperature for this person is different from $98.6°F$.

C.59   (a) Placebo-controlled means participants are given a treatment that is as close to the real treatment as feasible – in this case, dark chocolate which has had the flavonoids removed.

(b) This is a hypothesis test for a difference in means. Using μ_C for the mean increase in flow-mediated dilation for people eating dark chocolate every day and μ_N for the mean increase in flow-mediated dilation for people eating a dark chocolate substitute each day, the hypotheses for the test are

$$H_0 : \quad \mu_C = \mu_N$$
$$H_a : \quad \mu_C > \mu_N$$

The relevant statistic for this test is $\bar{x}_C - \bar{x}_N$, the difference in means for the two samples. The relevant null parameter is zero, since from the null hypothesis we have $\mu_C - \mu_N = 0$. The t-test statistic is:

$$t = \frac{\text{Sample statistic} - \text{Null parameter}}{SE} = \frac{(\bar{x}_C - \bar{x}_N) - 0}{\sqrt{\frac{s_C^2}{n_C} + \frac{s_N^2}{n_N}}} = \frac{1.3 - (-0.96)}{\sqrt{\frac{2.32^2}{11} + \frac{1.58^2}{10}}} = 2.63.$$

This is an upper-tail test, so using the t-distribution with df = 9, we see that the p-value is 0.014. This is a reasonably small p-value, so we reject H_0, and conclude that there is evidence that dark chocolate improves vascular health. The results are significant at a 5% level but not at a 1% level. (This is not surprising given the very small sample sizes. The fact that the results are significant at all is pretty impressive.)

(c) Yes, the results are significant and come from a randomized experiment.

C.61   (a) We test $H_0 : \mu_1 = \mu_2$ vs $H_a : \mu_1 > \mu_2$ where μ_1 and μ_2 are the mean AAMP scores for athletes using Tribulus and those not using it, respectively. We proceed with some caution since the sample sizes are rather small and the article did not include any information about possible outliers or skewness in the data. The relevant test statistic is

$$t = \frac{\bar{x}_1 - \bar{x}_2}{\sqrt{\frac{s_1^2}{n_1} + \frac{s_2^2}{n_2}}} = \frac{1305.6 - 1255.9}{\sqrt{\frac{177.3^2}{20} + \frac{66.8^2}{12}}} = 1.13$$

Using a t-distribution with 11 df, we find that the p-value in the upper tail is 0.141. We do not have sufficient evidence to conclude that athletes using Tribulus have a higher mean AAMP score than those not using the supplement.

(b) To compare the mean AAMP using measurements on the same subjects (the 20 athletes in the experimental group) before and after using the Tribulus supplement we would need the paired data to find the difference for each subject. There is no way to recover the standard deviation of the differences (and complete the test) by knowing just the mean and standard deviation for all 20 participants before and after using the supplement.

C.63 (a) For Mid-Atlantic houses the count in both groups is greater than 10, so the normal distribution is appropriate. For the California houses we see only 7 big houses, so the normal distribution may not be appropriate.

(b) We can use the test based on the normal distribution. Using p to represent the proportion of Mid-Atlantic homes larger than the national average, we are testing $H_0 : p = 0.25$ vs $H_a : p < 0.25$. The sample proportion for big homes in the Mid-Atlantic states is $\hat{p} = 16/90 = 0.178$, We calculate the test statistic

$$z = \frac{\hat{p} - p_0}{\sqrt{\frac{p_0(1-p_0)}{n}}} = \frac{0.178 - 0.25}{\sqrt{\frac{0.25(1-0.25)}{90}}} = -1.58$$

The area in a standard normal distribution below $z = -1.58$ gives p-value $= 0.057$. This is not (quite) small enough to reject H_0 at a 5% significance level. We do not have sufficient evidence to show that the proportion of big houses for sale in these Mid-Atlantic states is less than 25%.

(c) Because the CLT may not apply, we return to the methods of Chapter 4 and perform a randomization test to test the California proportion. The observed sample proportion is $\hat{p} = 7/30 = 0.233$. We use **StatKey** or other technology to generate a randomization distribution assuming $p = 0.25$, and find the proportion of simulated randomizations yielding sample proportions less than or equal to 0.233.

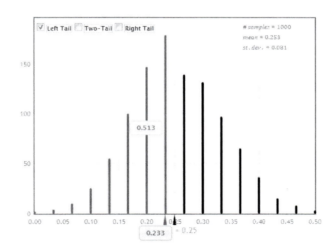

This proportion in this set of randomization samples is 0.513, so our p-value $= 0.513$. We fail to reject the null hypothesis, meaning we don't have sufficient evidence that the proportion of California houses considered big is less than 25%.

C.65 (a) For a 99% confidence interval with $df = 37$, we have $t^* = 2.72$. The confidence interval is

$$\overline{x} \;\pm\; t^* \cdot \frac{s}{\sqrt{n}}$$
$$111.7 \;\pm\; 2.72 \cdot \frac{144}{\sqrt{38}}$$
$$111.7 \;\pm\; 63.5$$
$$48.2 \;\text{to}\; 175.2$$

The best estimate for the mean time to infection for all kidney dialysis patients is 111.7 days with a margin of error of 63.5. We are 99% confident that the mean time to infection for all kidney dialysis patients is between 48.2 days and 175.2 days.

(b) Both 24 days and 165 days are reasonable values for individual patients. Both values lie well within one standard deviation ($s = 144$) of the estimated mean of 111.7 days. In fact, the actual data in the sample range from 2 days to 536 days. Remember that the confidence interval is an interval for the *mean* of the population – not a range for individual values.

(c) Since the confidence interval goes from 48.2 to 175.2, it would be implausible for the mean in the population to be as small as 24 days, but a mean of 165 days wold be a plausible value for the population.

C.67 This is a hypothesis test for a difference in means. Using μ_L for the average weight gain of mice with light at night and μ_D for the average weight gain of mice with darkness at night, the hypotheses for the test are:

$$H_0: \quad \mu_L = \mu_D$$
$$H_a: \quad \mu_L > \mu_D$$

The relevant statistic for this test is $\overline{x}_L - \overline{x}_D$, where $\overline{x}_L$ represents the mean weight gain of the sample mice in light and $\overline{x}_D$ represents the mean weight gain of the sample mice in darkness. The relevant null parameter is zero, since from the null hypothesis we have $\mu_L - \mu_D = 0$. The t-test statistic is:

$$t = \frac{\text{Sample statistic} - \text{Null parameter}}{SE} = \frac{(\overline{x}_L - \overline{x}_D) - 0}{\sqrt{\frac{s_L^2}{n_L} + \frac{s_D^2}{n_D}}} = \frac{9.4 - 5.9}{\sqrt{\frac{3.2^2}{19} + \frac{1.9^2}{8}}} = 3.52$$

This is a right tail test, so using the t-distribution with df $= 7$, we see that the p-value is 0.005. This is quite a small p-value so we reject H_0. There is strong evidence that mice with light at night gain significantly more weight on average than mice with darkness at night.

C.69 (a) We test $H_0: \mu_2 = 54$ vs $H_a: \mu_2 > 54$ where μ_2 is the mean score for all of the instructor's students on the CAOS posttest. The mean for the sample of 10 students is $\overline{x}_2 = 60.25$ with standard deviation $s_2 = 9.96$. The relevant t-statistic is

$$t = \frac{60.25 - 54.0}{9.96/\sqrt{10}} = 1.98$$

A dotplot of the posttest scores is relatively symmetric with no strong outliers so we use a t-distribution with 9 df to find the upper-tail p-value $= 0.040$. Since this p-value is less than 5%, we reject H_0 and have evidence that the mean score for this instructor's students on the CAOS posttest is higher than the benchmark mean of 54.0 points.

(b) We test $H_0: \mu_1 = 44.9$ vs $H_a: \mu_1 > 44.9$ where μ_1 is the mean score for all of the instructor's students on the CAOS pretest. The mean for the sample of 10 students is $\overline{x}_1 = 46.25$ with standard deviation $s_1 = 9.30$. The relevant t-statistic is

$$t = \frac{46.25 - 44.9}{9.30/\sqrt{10}} = 0.46$$

A dotplot of the pretest scores is relatively symmetric with no strong outliers so we use a t-distribution with 9 df to find the upper-tail p-value $= 0.328$. Since this p-value is quite large, we do not find sufficient evidence to reject H_0. The mean score for this instructor's students on the CAOS pretest is not significantly more than the benchmark mean of 44.9 points.

(c) We test $H_0 : \mu_d = 9.1$ vs $H_a : \mu_d > 9.1$ where μ_d is the mean improvement (*Posttest − Pretest*) for all of the instructor's students on the CAOS exams. To test this hypothesis, we need to compute the differences for the 10 students in the sample.

$$17.5, \quad 5, \quad 7.5, \quad 15, \quad 5, \quad 25, \quad 15, \quad 17.5, \quad 15, \quad 17.5$$

The mean improvement for the sample of 10 students is $\bar{x}_d = 14.0$ with standard deviation $s_d = 6.37$. The relevant t-statistic is

$$t = \frac{14.0 - 9.1}{6.37/\sqrt{10}} = 2.43$$

A dotplot of the improvement differences is relatively symmetric with no strong outliers so we use a t-distribution with 9 df to find the upper-tail p-value = 0.019. Since this p-value is less than 5%, we reject H_0 and have evidence that the mean improvement for this instructor's students on the CAOS exam is higher than the benchmark mean of 9.1 points.

C.71 Here is some computer output for doing both a confidence interval for $\mu_f - \mu_m$ and a test of $H_0 : \mu_f = \mu_m$ vs $H_a : \mu_f \neq \mu_m$. (Note: You could also use the summary statistics to compute the details of the test by hand.)

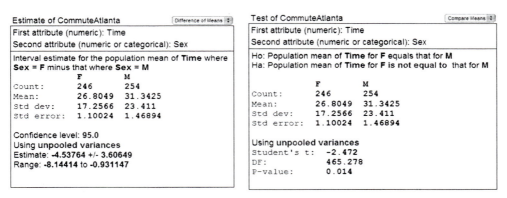

Contrary to the St. Louis commute results, the p-value in this case is small (0.014) so we have evidence of a difference in mean commute time by gender in Atlanta. From the confidence interval, we are 95% sure that the mean commute time for females in Atlanta is between 8.14 and 0.93 minutes less than the mean commute time for males in Atlanta.

C.73 (a) If we let μ_h and μ_w denote the mean age at marriage for husbands and wives respectively, the relevant hypotheses are $H_0 : \mu_h = \mu_w$ vs $H_a : \mu_h > \mu_w$. Since these are paired data, collected from $n = 105$ couples, we could also use the hypotheses $H_0 : \mu_d = 0$ vs $H_a : \mu_d > 0$ where μ_d is the mean difference in age, *Husband − Wife*.

If needed we use technology to compute the difference in age for each couple and then run a single sample t-test to see if the mean difference is greater than zero. For this sample husbands are older by an average of $\bar{x}_d = 2.83$ years with a standard deviation of $s_d = 5.00$. The t-statistic is 5.80 and the p-value for the upper-tail test is essentially zero. This gives very strong evidence that on average the husbands are older.

(b) Using the data in **MarriageAges** we find that the husband is older than the wife in 75 of the 105 marriages or 71.4% of the time. To see if this provides evidence that the proportion is bigger than 0.5 for the entire population we use technology to test $H_0 : p = 0.5$ vs $H_a : p > 0.5$. The z-statistic for the test is $z = 4.39$ and the p-value is essentially zero. This gives very strong evidence that the husband is older than the wife in more than 50% of recently married couples in St. Lawrence County.

(c) Using technology with the differences from part (a) we find that a 95% confidence interval for the difference in mean ages goes from 1.87 to 3.80 years. We are 95% sure that, on average, husbands are between 1.87 and 3.80 years older than their wives.

Using the fact that the husband was older in 75 out of 105 couples sampled, technology indicates that a 95% confidence interval for the proportion goes from 0.618 to 0.798. We are 95% sure that the husband is older than his wife in between 61.8% and 79.8% of newly married couples in St. Lawrence County.

C.75 These data are paired, so we use only the *Difference* variable to do paired difference in means inference. Using a software package and the variable *Difference* in the dataset **TrafficFlow**, we obtain the following output from one statistics package:

```
One-Sample T: Difference
Test of mu = 0 vs not = 0

Variable     N   Mean   StDev  SE Mean      95% CI         T      P
Difference  24  61.00  15.19     3.10  (54.59, 67.41)  19.68  0.000
```

The mean difference is 61.0 minutes. Since the differences are positive, and the subtraction went as $Difference = Timed - Flexible$, we see that the timed method has higher mean times in this sample compared to the flexible times. We see that a 95% confidence interval is 54.59 to 67.41. The flexible method reduces delay time by an average of between 54.59 and 67.41 minutes in simulations. We see from the p-value of 0.000 that the difference is significant: the flexible system is clearly better than the timed system. In order for this inference to be valid, note that we must assume that the simulations run by the engineers are a representative sample of all types of traffic flow on these streets.

Section 7.1 Solutions

7.1 The expected count in each category is $n \cdot p_i = 500(0.25) = 125$. See the table.

Category	1	2	3	4
Expected count	125	125	125	125

7.3 The expected count in category A is $n \cdot p_A = 200(0.50) = 100$. The expected count in category B is $n \cdot p_B = 200(0.25) = 50$. The expected count in category C is $n \cdot p_C = 200(0.25) = 50$. See the table.

Category	A	B	C
Expected count	100	50	50

7.5 We calculate the chi-square statistic using the observed and expected counts.

$$\chi^2 = \sum \frac{(observed - expected)^2}{expected}$$
$$= \frac{(35-40)^2}{40} + \frac{(32-40)^2}{40} + \frac{(53-40)^2}{40}$$
$$= 0.625 + 1.6 + 4.225$$
$$= 6.45$$

There are three categories (A, B, and C) for this categorical variable, so we use a chi-square distribution with degrees of freedom equal to 2. The p-value is the area in the upper tail, which we see is 0.0398.

7.7 We calculate the chi-square statistic using the observed and expected counts.

$$\chi^2 = \sum \frac{(observed - expected)^2}{expected}$$
$$= \frac{(132-160)^2}{160} + \frac{(181-160)^2}{160} + \frac{(45-40)^2}{40} + \frac{(42-40)^2}{40}$$
$$= 4.90 + 2.76 + 0.63 + 0.10$$
$$= 8.38$$

There are four categories (A, B, C, and D) for this categorical variable, so we use a chi-square distribution with degrees of freedom equal to 3. The p-value is the area in the upper tail, which we see is 0.039.

7.9 (a) The sample size is $n = 160$ and the hypothesized proportion is $p_b = 0.25$, so the expected count is $n \cdot p_b = 160(0.25) = 40$.

(b) For the "B" cell we have

$$\frac{(observed - expected)^2}{expected} = \frac{(36-40)^2}{40} = 0.4$$

(c) The table has $k = 4$ cells, so the chi-square distribution has $4 - 1 = 3$ degrees of freedom.

7.11 (a) Add the counts in the table to find the sample size is $n = 210+732+396+125+213+324 = 2000$. The hypothesized proportion is $p_b = 0.35$, so the expected count is $n \cdot p_b = 2000(0.35) = 700$.

(b) For the "B" cell we have

$$\frac{(\text{observed} - \text{expected})^2}{\text{expected}} = \frac{(732 - 700)^2}{700} = 1.46$$

(c) The table has $k = 6$ cells, so the chi-square distribution has $6 - 1 = 5$ degrees of freedom.

7.13 (a) Let p_g, p_o, p_p, p_r, and p_y be the proportion of people who choose each of the respective flavors. If all flavors are equally popular ($1/5$ each) the hypotheses are

$$\begin{aligned} H_0 : & \quad p_g = p_o = p_p = p_r = p_y = 0.2 \\ H_a : & \quad \text{Some } p_i \neq 0.2 \end{aligned}$$

(b) If they were equally popular we would have $66(1/5) = 13.2$ people in each category.

(c) Since we have 5 categories we have 4 degrees of freedom.

(d) We calculate the test statistic

$$\begin{aligned} \chi^2 &= \frac{(18 - 13.2)^2}{13.2} + \frac{(9 - 13.2)^2}{13.2} + \frac{(15 - 13.2)^2}{13.2} + \frac{(13 - 13.2)^2}{13.2} + \frac{(11 - 13.2)^2}{13.2} \\ &= 1.75 + 1.34 + 0.25 + 0.00 + 0.37 \\ &= 3.70 \end{aligned}$$

(e) The test statistic 3.70 compared to a chi-square distribution with 4 degrees of freedom yields a p-value of 0.449. We fail to reject the null hypothesis, meaning the data doesn't provide significant evidence that some skittle flavors are more popular than others.

7.15 (a) Since there are five groups and we are assuming the groups are equally likely, the proportion in each group should be $1/5$ or 0.2. The hypotheses for the test are:

$$\begin{aligned} H_0 : & \quad p_a = p_2 = p_3 = p_4 = p_5 = 0.2 \\ H_a : & \quad \text{Some } p_i \text{ is not } 0.2 \end{aligned}$$

where p_1 represents the proportion of social networking site users in the 18-22 age group, p_2 represents the proportion in the 23-35 age group, and so on. We have $n = 975$ and $p_i = 0.2$ for each group, so the expected count if all age groups are equally likely is $975(0.2) = 195$. The expected counts are all 195, as shown in the table.

Age	18-22	23-35	36-49	50-65	65+
Expected count	195	195	195	195	195

We calculate the chi-square statistic using the observed and expected counts.

$$\begin{aligned} \chi^2 &= \sum \frac{(observed - expected)^2}{expected} \\ &= \frac{(156 - 195)^2}{195} + \frac{(312 - 195)^2}{195} + \frac{(253 - 195)^2}{195} + \frac{(195 - 195)^2}{195} + \frac{(59 - 195)^2}{195} \\ &= 7.8 + 70.2 + 17.3 + 0 + 94.9 \\ &= 190.2 \end{aligned}$$

There are five age groups, so we use a chi-square distribution with degrees of freedom equal to 4. The p-value is the area in the upper tail, which we see is essentially zero. There is strong evidence that users of social networking sites are not equally distributed among these age groups.

(b) The largest contributor to the sum for the chi-square test statistic is 94.9 from the 65+ age group, where the observed count is significantly below the expected count. Many fewer senior citizens use a social networking site than we would expect if users were evenly distributed by age.

7.17 Let p_1, p_2, p_3, and p_4 be the proportion of hockey players born in the 1st, 2nd, 3rd, and 4th quarter of the year, respectively. We are testing

$H_0 : p_1 = 0.237, \; p_2 = 0.259, \; p_3 = 0.259$ and $p_4 = 0.245$
H_a: Some p_i is not specified as in H_0

The total sample size is $n = 147 + 110 + 52 + 50 = 359$. The expected counts are $359(0.237) = 85$ for Qtr 1, $359(0.259) = 93$ for Qtr 2, $359(0.259) = 93$ for Qtr 3, and $359(0.245) = 88$ for Qtr 4. The chi-square statistic is

$$\chi^2 = \frac{(147 - 85)^2}{85} + \frac{(110 - 93)^2}{93} + \frac{(52 - 93)^2}{93} + \frac{(50 - 88)^2}{88} = 82.6.$$

We use the chi-square distribution with $4 - 1 = 3$ degrees of freedom, which gives a very small p-value that is essentially zero. This is strong evidence that the distribution of birthdates for OHL hockey players differs significantly from the national proportions.

7.19 (a) A χ^2 goodness-of-fit test was most likely done.

(b) Since the results are given as statistically significant, the χ^2-statistic is likely to be large.

(c) Since the results are given as statistically significant, the p-value is likely to be small.

(d) The categorical variable appears to record the number of deaths due to medication errors in different months at hospitals.

(e) The cell giving the number of deaths in July appears to contribute the most to the χ^2- statistic.

(f) In July, the observed count is probably much higher than the expected count.

7.21 (a) There are 6 actors and we are testing for a difference in popularity. The null hypothesis is that each of the proportions is 1/6 while the alternative hypothesis is that at least one of the proportions is not 1/6. The sample size is $98 + 5 + 23 + 9 + 25 + 51 = 211$, so the expected count for each actor is

$$\text{Expected count for each actor } = n \cdot p_i = 211(1/6) = 35.2$$

The chi-square test statistic calculated using the observed data and these expected counts

$$
\begin{aligned}
\chi^2 &= \sum \frac{(observed - expected)^2}{expected} \\
&= \frac{(98 - 35.2)^2}{35.2} + \frac{(5 - 35.2)^2}{35.2} + \frac{(23 - 35.2)^2}{35.2} + \frac{(9 - 35.2)^2}{35.2} + \frac{(25 - 35.2)^2}{35.2} + \frac{(51 - 35.2)^2}{35.2} \\
&= 112.3 + 25.9 + 4.2 + 19.5 + 3.0 + 7.1 \\
&= 172.0
\end{aligned}
$$

This chi-square statistic gives a very small p-value of essentially zero when compared to a chi-square distribution with 5 degrees of freedom. There is strong evidence of a difference in the popularity of the James Bond actors.

(b) If we eliminate one actor, the null hypothesis is that each of the proportions is $1/5$ while the alternative hypothesis is that at least one of the proportions is not $1/5$. The sample size without the 5 people who selected George Lazenby is $98 + 23 + 9 + 25 + 51 = 206$, so the expected count for each actor is

$$\text{Expected count for each actor} = n \cdot p_i = 206(1/5) = 41.2$$

The chi-square statistic calculated using the observed data (without Lazenby) and these expected counts is

$$
\begin{aligned}
\chi^2 &= \sum \frac{(observed - expected)^2}{expected} \\
&= \frac{(98 - 41.2)^2}{41.2} + \frac{(23 - 41.2)^2}{41.2} + \frac{(9 - 41.2)^2}{41.2} + \frac{(25 - 41.2)^2}{41.2} + \frac{(51 - 41.2)^2}{41.2} \\
&= 78.3 + 8.0 + 25.2 + 6.4 + 2.3 \\
&= 120.2
\end{aligned}
$$

This is still a very large χ^2-statistic and we again have a p-value of essentially zero when it is compared to a chi-square with 4 degrees of freedom. Even with the Lazenby data omitted, we sill find substantial differences in he proportions of fans who choose the different James Bond actors.

(c) No, we should not generalize the results from this online survey to a population of all movie watchers. This poll was a volunteer poll completed by people visiting a James Bond fan site. This is definitely not a random sample of the movie watching population, and could easily be biased. The best inference we could hope for is to generalize to people who visit a James Bond fan website and who participate in online polls.

7.23 (a) We see in the bottom row of the output that $n = 436$. (We could also add the observed counts.)

(b) We see in the output that the observed value for RR is 130 and the expected count is 109.

(c) The contribution to the χ^2 statistic is the highest for those with the XX variant, which contributes 7.716. For this category, the observed count (80) is less than expected (109).

(d) We see in the bottom row of the output that df=2. (We could also compute df using 3 categories minus 1.)

(e) We see in the bottom row of the output that the p-value is 0.002. This is a small p-value and provides evidence that at least one of the proportions of the three variants differ from the values of 0.25, 0.5, and 0.25, respectively.

7.25 According to Benford's Law the hypotheses are

$H_0 :$ $p_1 = 0.301, \ p_2 = 0.176, \ p_3 = 0.125, \ p_4 = 0.097, \ p_5 = 0.079,$
$\qquad p_6 = 0.067, \ p_7 = 0.058, \ p_8 = 0.051, \ p_9 = 0.046$
$H_a :$ At least one of the proportions is different from Benford's law

Here is a table of observed counts for the addresses and expected counts using the Benford proportions and a sample size of 1188.

Digit	1	2	3	4	5	6	7	8	9
Observed	345	197	170	126	101	72	69	51	57
Expected	357.6	209.2	148.4	115.1	94.1	79.5	68.9	60.8	54.4

The value of the chi-square statistic is

$$\chi^2 = \sum \frac{(observed - expected)^2}{expected} = \frac{(345 - 357.6)^2}{357.6} + \cdots + \frac{(57 - 54.4)^2}{54.4} = 8.24$$

We find the p-value $= 0.41$ using the upper tail beyond 8.24 of a chi-square distribution with $9 - 1 = 8$ degrees of freedom. This is not a small p-value so we do not have convincing evidence that the first digits of street addresses in a phone book do not follow Benford's law. Note that this doesn't *prove* Benford's law in this situation; we only have a lack of evidence against it.

7.27 For a pair of fair dice, the proportion for each of the possible sums are shown in the table below, along with the observed and expected counts from the sample of 180 rolls.

Dice sum	2	3	4	5	6	7	8	9	10	11	12
Null proportion	1/36	2/36	3/36	4/36	5/36	6/36	5/36	4/36	3/36	2/36	1/36
Observed count	5	11	16	13	26	34	19	20	16	13	7
(Expected count)	(5)	(10)	(15)	(20)	(25)	(30)	(25)	(20)	(15)	(10)	(5)

For the null proportions, there are $6 \cdot 6 = 36$ possible results for the two dice and only one $(1,1)$ gives a sum of two while there are six ways to get a sum of seven, $(1,6), (2,5), (3,4), (4,3), (5,2), (6,1)$.

The null hypothesis is that the proportion for each sum is as shown in the table above and the alternative is that one or more of these proportions is inaccurate. We find the expected counts in the table by multiplying each of these proportions by the sample size of $n = 180$. The chi-square statistic is computed as

$$\chi^2 = \frac{(5 - 5)^2}{5} + \frac{(11 - 10)^2}{10} + \frac{(16 - 15)^2}{15} + \cdots + \frac{(13 - 10)^2}{10} + \frac{(7 - 5)^2}{5} = 6.40$$

Comparing $\chi^2 = 6.40$ to the upper-tail of a chi-square distribution with 10 degrees of freedom yields a p-value of 0.781. We have very little to no evidence that this author can roll some numbers more often than should naturally appear by random chance.

7.29 The hypotheses are
$H_0 : p_0 = p_1 = p_2 = \cdots = p_9 = 0.10$
$H_a :$ Some $p_i \neq 0.10$

Here are the observed and expected counts for the digits in $SSN8$.

Digit	0	1	2	3	4	5	6	7	8	9
Observed	13	14	16	13	14	15	17	15	15	18
Expected	15	15	15	15	15	15	15	15	15	15

Using technology we obtain the following output for this test

```
X-squared = 1.6, df = 9, p-value = 0.9963
```

This is a very large p-value so we have no substantial evidence that the eighth digits of social security numbers are not random.

Here are the observed and expected counts for the digits in $SSN9$, the last digit.

Digit	0	1	2	3	4	5	6	7	8	9
Observed	16	12	17	15	12	10	15	27	15	11
Expected	15	15	15	15	15	15	15	15	15	15

Using technology we obtain the following output for this test

```
X-squared = 13.8667, df = 9, p-value = 0.1271
```

This is not a small p-value so we lack sufficient evidence to conclude the last digits of social security numbers are not random.

Section 7.2 Solutions

7.31 For the (B,E) cell we have

$$\text{Expected count} = \frac{\text{B row total} \cdot \text{E column total}}{n} = \frac{330 \cdot 160}{600} = 88$$

The contribution to the χ^2-statistic from the (B,E) cell is $\dfrac{(89 - 88)^2}{88} = 0.011$.

7.33

We need to find a row total for Group 2, $1180 + 320 = 1500$, and the column total for No, $280 + 320 = 600$. Since the row total for Group 1 is $720 + 280 = 1000$, the overall sample size is $n = 1000 + 1500 = 2500$. For the (Group 2, No) cell we have

$$\text{Expected count} = \frac{1500 \cdot 600}{2500} = 360$$

The contribution to the χ^2-statistic from the (Group 2, No) cell is $\dfrac{(320 - 360)^2}{360} = 4.44$.

7.35 This is a 3×4 table so we have $(3 - 1) \cdot (4 - 1) = 6$ degrees of freedom. Also, if we eliminate the last row and last column (ignoring the totals) there are six cells remaining.

7.37 This is a 2×2 table so we have $(2 - 1) \cdot (2 - 1) = 1$ degree of freedom. If the row and column totals are known, we only need the value for any one cell to be able to fill in the rest of the table.

7.39 The null hypothesis is that the attitude about one true love is not related to one's educational level and the alternative hypothesis is that the two variables are related in some way. We compute expected counts for all 9 cells. For example, for the (Agree, Some) cell, we have

$$Expected = \frac{735 \cdot 668}{2625} = 187.0$$

Computing all the expected counts in the same way, we find the expected counts shown in the table. Note that all cell counts are large enough to use a χ^2-test.

	HS	Some	College
Agree	263.2	187.0	284.8
Disagree	648.9	461.1	702.0
Don't know	27.9	19.8	30.2

We compute $(observed - expected)^2/expected$ for each cell. The results are shown in the next table.

	HS	Some	College
Agree	37.71	0.65	27.69
Disagree	13.02	0.05	10.78
Don't know	2.24	1.94	0.11

Adding up all of these contributions, we obtain the χ^2-statistic 93.7. This is a very large test statistic, and the p-value from a chi-square distribution with $df = 4$ is essentially zero. There is very strong evidence of an association between education level and how one feels about whether we all have one true love.

We see that the largest contribution to the χ^2-statistic is in those who agree, with more high school educated people than expected agreeing and fewer college educated people than expected agreeing. It appears that the greater the amount of education, the less likely a person is to agree that we each have exactly one true love.

7.41 (a) Based on the given counts, here is the two-way table, with totals included.

	Relapse	No relapse	Total
Desipramine	10	14	24
Lithium	18	6	24
Placebo	20	4	24
Total	48	24	72

(b) The expected count for the (Desipramine, Relapse) cell is $(48 \cdot 24)/72 = 16$. All expected counts are shown in the table. Since the sample size is the same for each group, we see that the expected counts are the same from row to row, matching the null hypothesis that the treatment drug doesn't matter. Since all the expected counts are greater than 5, a chi-square test is appropriate.

	Relapse	No relapse
Desipramine	16	8
Lithium	16	8
Placebo	16	8

(c) The null hypothesis is that the drug does not affect the likelihood of a relapse, and the alternative hypothesis is that the drug does matter in the chances of recovery. We use the observed and expected counts to find the χ^2 statistic is

$$\chi^2 = \frac{(10-16)^2}{16} + \frac{(14-8)^2}{8} + \frac{(18-16)^2}{16} + \frac{(6-8)^2}{8} + \frac{(20-16)^2}{16} + \frac{(4-8)^2}{8} = 10.5$$

Using the χ^2 distribution with $df = 2$, we find a p-value of 0.005. There is strong evidence that the drug used is related to the likelihood of a relapse.

(d) Desipramine appears to be significantly more effective than lithium or a placebo. Yes, we can conclude that the drug affects the likelihood of successful recovery, since the results come from a randomized experiment.

7.43 (a) The hypotheses are
H_0 : Skittles choice does not depend on method of choosing (color vs flavor)
H_a : Skittles choice depends on method of choosing

(b) The expected counts under H_0 are

	Green(Lime)	Orange	Purple(Grape)	Red(Strawberry)	Yellow(Lemon)
Color	13.0	10.5	14.3	19.8	8.4
Flavor	18.0	14.5	19.7	27.2	11.6

(c) All of the expected counts are larger then 5, so we can use a chi-square test.

(d) We have $(5-1)(2-1) = 4$ degrees of freedom.

(e) The chi-square statistic is

$$\chi^2 = \frac{(18-13.0)^2}{13.0} + \frac{(9-10.5)^2}{10.5} + \cdots + \frac{(9-11.6)^2}{11.6} = 9.07$$

(f) Comparing our test statistic to a chi-square distribution with 4 degrees of freedom we get a p-value of 0.059. This is right on the border, so we see weak evidence that choosing flavor vs color might affect the choices, but not enough to reject the null hypothesis if we are using a 5% level.

7.45 The hypotheses are

H_0 : Age distribution is the same in 2008 and 2010
H_a : Age distribution is different between 2008 and 2010

We compute expected counts for all 8 cells. For example, for the 18-22 year olds in 2008, we have

$$\text{Expected count for ages 18-22 in 2008} = \frac{290 \cdot 495}{1442} = 99.5.$$

Computing all the expected counts in the same way, we find the expected counts shown in the table.

↓Age/Year→	2008	2010
18-22	99.5	190.5
23-35	171.6	328.4
36-49	121.5	232.5
50+	102.3	195.7

Notice that all expected counts are over 5 so we can use a chi-square distribution. We compute the contribution to the χ^2-statistic, $(observed - expected)^2/expected$, for each cell. The results are shown in the next table.

↓Age/Year→	2008	2010
18-22	14.9	7.8
23-35	3.7	2.0
36-49	1.5	0.8
50+	24.7	12.9

Adding up all of these contributions, we obtain the χ^2-statistic 68.3. This test statistic is very large. Using a chi-square distribution with $df = (4-1) \cdot (2-1) = 3$, we see that the p-value is essentially zero. We have very strong evidence that the age distribution changed during these two years. We see that the biggest contribution to the chi-square statistic is from the 50 and older groups in the year 2008, where the number of social network site users was below expected (based on H_0) in 2008.

7.47 The null hypothesis is that frequency of status updates on Facebook is not different for males and females. The alternative hypothesis is that frequency of status updates is related to gender. We compute expected counts for all 10 cells. For example, for the males who update their status every day, we have

$$Expected = \frac{130 \cdot 386}{877} = 57.2.$$

Computing all the expected counts in the same way, we find the expected counts shown in the following table.

↓Status/Gender→	Male	Female
Every day	57.2	72.8
3-5 days/week	46.2	58.8
1-2 days/week	65.6	83.4
Every few weeks	68.7	87.3
Less often	148.3	188.7

Notice that all expected counts are over 5 so we can use a chi-square distribution. We compute the contribution to the χ^2-statistic, $(observed - expected)^2/expected$, for each cell and show the results in the next table.

↓Status/Gender→	Male	Female
Every day	4.05	3.18
3-5 days/week	0.0	0.0
1-2 days/week	0.33	0.23
Every few weeks	1.01	0.80
Less often	0.05	0.04

Adding up all of these contributions, we obtain the χ^2-statistic 9.65. Using a chi-square distribution with $df = (5 - 1) \cdot (2 - 1) = 4$, we see that the p-value is 0.047. This is very close to a 5% cutoff. At a 5% level, there is some evidence of a possible relationship between gender and frequency of status updates on Facebook. Females appear more likely to update their status at least once a day.

7.49 (a) The expected count in the (Endurance, XX) cell is 34.75, and the contribution of this cell to the chi-square statistic is 3.645. We find the expected count using

$$\frac{\text{Endurance row total} \cdot XX \text{ column total}}{\text{Sample size}} = \frac{194 \cdot 132}{737} = 34.75$$

The contribution to the chi-square statistic is

$$\frac{(observed - expected)^2}{expected} = \frac{(46 - 34.75)^2}{34.75} = 3.642$$

which is the same (up to round-off) as the computer output.

(b) We see in the bottom row of the computer output that "DF = 4". Since the two-way table has 3 rows and 3 columns, we have $df = (3 - 1) \cdot (3 - 1) = 4$, as given.

(c) We see in the bottom row of the output that the chi-square test statistic is 24.805 and the p-value is 0.000. There is strong evidence that the distribution of genotypes for this gene is different between sprinters, endurance athletes, and non-athletes.

(d) The (Sprint, XX) cell contributes the most, 9.043, to the χ^2-statistic. The observed count (6) is substantially less than the expected count (19.16). Sprinters are not likely to have this genotype.

(e) The genotype RR is most over-represented in sprinters (53 compared to an expected count of 35.28). The genotype XX is most over-represented in endurance athletes (46 compared to an expected count of 34.75).

7.51 The p-value is 0.592, which is a large p-value. The sample provides no evidence at all that genotype distribution is different between males and females. Gender does not appear to be associated with whether or not one has the "sprinting gene".

7.53 The null hypothesis is that the two variables are related and the alternative hypothesis is that the two variables are not related. The output from one statistics package (Minitab) is given. Output from other packages may look different but will give the same (or similar) chi-square statistic and p-value. We see that the p-value is 0.004 so there is a significant association between these two variables. The largest contribution to the chi-square statistic is from the males who do not take vitamins, and we see that males are less likely to take vitamins than expected.

```
Rows: Gender   Columns: VitaminUse

              No  Occasional  Regular

Female        87          77      109
            96.20       71.07   105.73
           0.8798      0.4954   0.1009

Male          24           5       13
            14.80       10.93    16.27
           5.7189      3.2199   0.6560

Cell Contents:      Count
                    Expected count
                    Contribution to Chi-square

Pearson Chi-Square = 11.071, DF = 2, P-Value = 0.004
```

Section 8.1 Solutions

8.1 Both datasets have the same group means of $\bar{x}_1 = 15$ and $\bar{x}_2 = 20$. However, there is so much variability within the groups for Dataset A that we really can't be sure of a difference. In Dataset B, since there is so much less variability between the groups, it seems obvious that the groups come from different populations. (Think of it this way: if the next number we saw was a 20, would we know which group it belongs to in Dataset A? No. Would we know which group it belongs to in Dataset B? Yes!) We have more convincing evidence for a difference between group 1 and group 2 in Dataset B, since the variability within groups is so much less.

8.3 The scales are the same and it appears that the variability is about the same for both datasets. However the means appear to be much farther apart in Dataset A, so we have more convincing evidence for a difference in population means in Dataset A.

8.5 The scales are the same and it appears that the sample means (about 15 and 25) are the same for the two datasets. However, there is much more variability within the groups in Dataset A, so have more convincing evidence for a difference in population means in Dataset B.

8.7 Since there are three groups, the degrees of freedom for the groups is $3 - 1 = 2$. Since the total number of data values is 15, the total degrees of freedom is $15 - 1 = 14$. This leaves 12 degrees of freedom for the Error (or use $n - \#groups = 15 - 3 = 12$). The Mean Squares are found by dividing each SS by its df, so we compute

$$MSG = 120/2 = 60 \quad \text{(for Groups)} \quad \text{and} \quad MSE = 282/12 = 23.5 \quad \text{(for Error)}$$

The F-statistic is the ratio of the mean square values, we have

$$F = \frac{MSG}{MSE} = \frac{60}{23.5} = 2.55$$

The completed table is shown below.

Source	df	SS	MS	F-statistic
Groups	2	120	60	2.55
Error	12	282	23.5	
Total	14	402		

8.9 Since there are three groups, degrees of freedom for the groups is $3 - 1 = 2$. Since the total number of data values is $10 + 8 + 11 = 29$, the total degrees of freedom is $29 - 1 = 28$. This leaves 26 degrees of freedom for the Error (or use $n - \#groups = 29 - 3 = 26$). We also find the missing sum of squares for Error by subtraction, $SSE = SSTotal - SSG = 1380 - 80 = 1300$. The Mean Squares are found by dividing each SS by its df, so we compute

$$MSG = 80/2 = 40 \quad \text{(for Groups)} \quad \text{and} \quad MSE = 1300/26 = 50 \quad \text{(for Error)}$$

The F-statistic is the ratio of the mean square values, we have

$$F = \frac{MSG}{MSE} = \frac{40}{50} = 0.8$$

The completed table is shown below.

Source	df	SS	MS	F-statistic
Groups	2	80	40	0.8
Error	26	1300	50	
Total	28	1380		

8.11 (a) Since the degrees of freedom for the groups is 3, the number of groups is 4.

(b) The hypotheses are

$$H_0: \quad \mu_1 = \mu_2 = \mu_3 = \mu_4$$
$$H_a: \quad \text{Some } \mu_i \neq \mu_j$$

(c) Using 3 and 16 for the degrees of freedom with the F-distribution, we see the upper-tail area beyond F=1.60 gives a p-value of 0.229.

(d) We do not reject H_0. We do not find convincing evidence for any differences between the population means.

8.13 (a) Since the degrees of freedom for the groups is 2, the number of groups is 3.

(b) The hypotheses are

$$H_0: \quad \mu_1 = \mu_2 = \mu_3$$
$$H_a: \quad \text{Some } \mu_i \neq \mu_j$$

(c) Using 2 and 27 for the degrees of freedom with the F-distribution, we see the upper-tail area beyond F=8.60 gives a p-value of 0.0013.

(d) We reject H_0. We find strong evidence for differences among the population means.

8.15 (a) One variable is which group the girl is in, which is categorical. The other variable is the change in cortisol level, which is quantitative.

(b) This is an experiment, since the researchers assigned the girls to the different groups.

(c) The hypotheses are

$$H_0: \quad \mu_1 = \mu_2 = \mu_3 = \mu_4$$
$$H_a: \quad \text{Some } \mu_i \neq \mu_j$$

where the four means represent the mean cortisol change after a stressful event for girls who talk to their mothers in person, who talk to their mothers on the phone, who text their mothers, and who have no contact with their mothers, respectively.

(d) Since the overall sample size is 68, total degrees of freedom is 67. Since there are four groups, the df for groups is 3. This leaves 64 degrees of freedom for the error (or use $68 - 4 = 64$).

(e) Since they found a significant difference in mean cortisol change between at least two of the groups, the F-statistic must be significant, which means its p-value is less than 0.05.

8.17 (a) Legs together with no lap pad has the largest temperature increase. Spreading the legs apart has the smallest temperature increase.

(b) Yes, the standard deviations are similar. The largest, $s_1 = 0.96$, is not more than twice the smallest, $s_3 = 0.66$.

(c) The null hypothesis is that the population mean temperature increases for the three conditions are all the same and the alternative hypothesis is that at least two of the means are different. We find the mean squares by dividing the sum of squares by the respective degrees of freedom ($df = 3 - 1 = 2$ for Groups, $df = 87 - 3 = 84$ for Error). The F-statistic is the ratio of the two mean squares.

$$F = \frac{MSG}{MSE} = \frac{6.85}{0.63} = 10.9$$

These calculations are summarized in the ANOVA table below.

Source	DF	SS	MS	F	P
Groups	2	13.7	6.85	10.9	0.0001
Error	84	53.2	0.63		
Total	86	66.9			

The p-value is the area above 10.9 in an F-distribution with numerator degrees of freedom 2 and denominator degrees of freedom 84. Using technology we see that the p-value=0.0001. We reject H_0 and find strong evidence that average temperature increase is not the same for these three conditions. It appears that spreading legs apart may be more effective at reducing the temperature increase.

8.19 (a) Yes, the control groups spend less time immobile than the groups that were stressed. The mean time immobile is smaller for each of the groups with "HC" than the means for any of the stressed groups with "SD". However, of the groups that were stressed, the mice that spent time in an enriched environment (EE) appear to spend much less time immobile (on average) than the mice in the other two stressed groups.

(b) The null hypothesis is that environment and prior stress do not affect mean amount of time immobile, while the alternative is that environment and prior stress do affect mean amount of time immobile. To construct the ANOVA table we compute

$$MSG = \frac{188464}{5} = 37692.8 \qquad MSE = \frac{197562}{42} = 4703.9 \qquad F = \frac{MSG}{MSE} = \frac{37692.8}{4703.9} = 8.0$$

We summarize these calculations in the ANOVA table below.

Source	DF	SS	MS	F	P
Light	5	188464	37692.8	8.0	0.00002
Error	42	197562	4703.9		
Total	47	659611			

From an F-distribution with 5 and 42 degrees of freedom, we see that the p-value is 0.00002. There is strong evidence that the average amount of time spent immobile is not the same for all six combinations of environment and stress. We will see in the next section how to tell where the differences lie.

8.21 The null hypothesis is that the mean change in pain threshold is the same regardless of pose struck, while the alternative hypothesis is that mean change in pain threshold is different between at least two of the types of poses. We find the sum of squares using the shortcut formulas at the end of this section. (The difference between $SSG + SSE$ and $SSTotal$ is due to rounding of the standard deviations.)

$$
\begin{aligned}
SSG &= 30(14.3 - 1.33)^2 + 29(-4.4 - 1.33)^2 + 30(-6.1 - 1.33)^2 = 7,654.9 \\
SSE &= 29(34.8)^2 + 28(31.9)^2 + 29(35.4)^2 = 99,954.9 \\
SSTotal &= 88(35.0)^2 = 107,800
\end{aligned}
$$

We fill in the rest of the ANOVA table by dividing each sum of square by its respective degrees of freedom and finding the ratio of the mean squares to compute the F-statistic.

$$
MSG = \frac{7654.9}{2} = 3827.5 \qquad MSE = \frac{99954.9}{86} = 1162.3 \qquad F = \frac{MSG}{MSE} = \frac{3827.5}{1162.3} = 3.29
$$

Using an F-distribution with 2 and 86 degrees of freedom, we find the area beyond $F = 3.29$ gives a p-value of 0.042. We reject H_0 and find evidence that the type of pose a person assumes is associated with change in mean pain threshold. The ANOVA table summarizing these calculations is shown below.

```
Source  DF    SS        MS      F      P
Light   2     7654.9    3827.5  3.29   0.042
Error   86    99954.9   1162.3
Total   88    107800
```

8.23 (a) Let μ_{genre} denote the mean Tomatometer score for each genre.

$$H_0 : \mu_{action} = \mu_{animation} = \mu_{comedy} = \mu_{drama} = \mu_{horror} = \mu_{romance} = \mu_{thriller}$$
$$H_a : \text{At least one } \mu_i \neq \mu_j$$

(b) Yes, the conditions are satisfied. Although some of the sample sizes are small, the normality condition does not appear to be strongly violated in any group. Also, the largest sample standard deviation, 31.57, is less than twice the smallest sample standard deviation, 18.06, so the condition of equal variability is satisfied.

(c) We find the df and sum of squares for error by subtracting the groups values from the total:

$$\text{Error df} = 130 - 6 = 124 \qquad SSE = 93,268.1 - 11,407.4 = 81,860.7$$

We complete the rest of the calculations in the ANOVA table below.

Source	d.f.	Sum of Sq.	Mean Square	F-statistic	p-value
Groups	6	11407.4	$\frac{11407.4}{6} = 1901.2$	$\frac{1901.2}{660.2} = 2.88$	0.012
Error	124	81860.7	$\frac{81860.7}{124} = 660.2$		
Total	130	93268.1			

(d) Yes. The p-value of 0.012 is less than $\alpha = 0.05$, so we have statistically significant evidence that mean Tomatometer score differs by movie genre.

(e) We may want to make inferences for all movies that have Tomatometer scores. We only have data on 2011 movies, which is not a random sample of all movies, and so may not be representative. We may not fully trust inferences to this larger population based on this sample.

8.25 (a) The null hypothesis is that the amount of light at night does not affect how much weight is gained. The alternative hypothesis is that the amount of light at night has some effect on mean weight gain.

(b) We see from the computer output that the F-statistic is 8.38 while the p-value is 0.002. This is a small p-value, so we reject H_0. There is evidence that mean weight gain in mice is influenced by the amount of light at night.

(c) Yes, there is an association between weight gain and light conditions at night. From the means of the groups, it appears that mice with more light at night tend to gain more weight.

(d) Yes, we can conclude that light at night causes weight gain (in mice), since the results are significant and come from a randomized experiment.

8.27 (a) The standard deviations are very different. In particular, the standard deviation for the LL sample ($s_{LL} = 1.31$) is more than double the standard deviation for the LD sample ($s_{LD} = 0.43$). This indicates that an ANOVA test may not be appropriate in this situation.

(b) A p-value of 0.652 from the randomization distribution is not small, so we would not reject a null hypothesis that the means are equal. There is not sufficient evidence to conclude that the mean amount consumed is different depending on the amount of light at night. Mice in different light at night conditions appear to eat the roughly similar amounts on average. Weight gain in mice with light at night is not a result of eating more food.

8.29 (a) For mice in this sample on a standard light/dark cycle, 36.0% of their food is consumed (on average) during the day and the other 64.0% is consumed at night. For mice with even just dim light at night, 55.5% of their food is consumed (on average) during the day and the other 44.5% is consumed at night.

(b) The p-value is 0.000 so there is very strong evidence that light at night influences when food is consumed by mice. Since the result comes from a randomized experiment, and the mice with more light average a higher percentage of food consumed during the day, we can conclude that light at night causes mice to eat a greater percentage of their food during the day. This raises the question: how strong is the association between time of calorie consumption and weight gain? That answer will have to wait until the next chapter.

8.31 We start by finding the mean and standard deviations for *Exercise* within each of the award groups and for the whole sample.

	Academy	Nobel	Olympic	Total
Mean	6.81	6.96	11.14	9.05
Std. Dev.	4.09	4.74	5.98	5.74
Sample size	31	148	182	361

The standard deviations are not too different, so the equal variance condition is reasonable. Side-by-side boxplots of exercise between the award groups show some right skewness, but the sample sizes are large so this shouldn't be a problem.

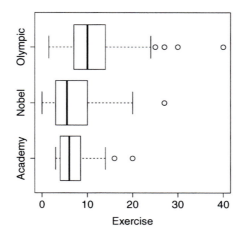

To test $H_0 : \mu_1 = \mu_2 = \mu_3$ vs $H_a :$ Some $\mu_i \neq \mu_j$ we use technology to produce an ANOVA table.

```
            Df   Sum Sq  Mean Sq  F value     Pr(>F)
Award        2   1597.9   798.96   27.861     0.0000
Residuals  358  10266.3    28.68
Total      360  11864.2
```

The p-value from the ANOVA (0.0000) is very small, so we have strong evidence that at least one of the award groups has a mean exercise rate that differs from at least one of the other groups.

Section 8.2 Solutions

8.33 Yes, the p-value (0.002) is very small so we have strong evidence that there is a difference in the population means between at least two of the groups.

8.35 We see in the output that the sample mean for group A is 10.2 with a sample size of 5. For a confidence interval for a mean after an analysis of variance test, we use $\sqrt{MSE}$ for the standard deviation, and we use error df for the degrees of freedom. For a 95% confidence interval with 12 degrees of freedom, we have $t^* = 2.18$. The confidence interval is

$$\overline{x}_A \quad \pm \quad t^* \cdot \frac{\sqrt{MSE}}{\sqrt{n_A}}$$

$$10.2 \quad \pm \quad 2.18 \cdot \frac{\sqrt{6.20}}{\sqrt{5}}$$

$$10.2 \quad \pm \quad 2.43$$

$$7.77 \quad \text{to} \quad 12.63$$

We are 95% confident that the population mean for group A is between 7.77 and 12.63.

8.37 We are testing $H_0 : \mu_A = \mu_C$ vs $H_a : \mu_A \neq \mu_C$. The test statistic is

$$t = \frac{\overline{x}_A - \overline{x}_C}{\sqrt{MSE\left(\frac{1}{n_A} + \frac{1}{n_C}\right)}} = \frac{10.2 - 10.8}{\sqrt{6.20\left(\frac{1}{5} + \frac{1}{5}\right)}} = -0.38$$

This is a two-tail test so the p-value is twice the area below -0.38 in a t-distribution with $df = 12$. We see that the p-value is $2(0.355) = 0.71$. This is a very large p-value and we do not find any convincing evidence of a difference in population means between groups A and C.

8.39 The pooled standard deviation is $\sqrt{MSE} = \sqrt{48.3} = 6.95$. We use the error degrees of freedom, which we see in the output is 20.

8.41 We see in the output that the sample mean for group C is 80.0 while the sample mean for group D is 69.33. Both sample sizes are 6. For a confidence interval for a difference in means after an analysis of variance test, we use $\sqrt{MSE}$ for both standard deviations, and we use the error df degrees of freedom. For a 95% confidence interval with 20 degrees of freedom, we have $t^* = 2.09$. The confidence interval is

$$(\overline{x}_C - \overline{x}_D) \quad \pm \quad t^* \cdot \sqrt{MSE\left(\frac{1}{n_C} + \frac{1}{n_D}\right)}$$

$$(80.00 - 69.33) \quad \pm \quad 2.09 \cdot \sqrt{48.3\left(\frac{1}{6} + \frac{1}{6}\right)}$$

$$10.67 \quad \pm \quad 8.39$$

$$2.28 \quad \text{to} \quad 19.06$$

We are 95% confident that the difference in the population means of group C and group D is between 2.28 and 19.06.

8.43 We are testing $H_0 : \mu_A = \mu_B$ vs $H_a : \mu_A \neq \mu_B$. The test statistic is

$$t = \frac{\overline{x}_A - \overline{x}_B}{\sqrt{MSE \left(\frac{1}{n_A} + \frac{1}{n_B} \right)}} = \frac{86.83 - 76.17}{\sqrt{48.3 \left(\frac{1}{6} + \frac{1}{6} \right)}} = 2.66$$

This is a two-tail test so the p-value is twice the area above 2.66 in a t-distribution with $df = 20$. We see that the p-value is $2(0.0075) = 0.015$. With a 5% significance level, we reject H_0 and find evidence of a difference in the population means between groups A and B.

8.45 To compare mean ant counts for peanut butter and ham & pickles the relevant hypotheses are $H_0 : \mu_2 = \mu_3$ vs $H_a : \mu_2 \neq \mu_3$. We compare the sample means, $\overline{x}_2 = 34.0$ and $\overline{x}_3 = 49.25$, and standardize using the SE for a difference in means after ANOVA.

$$t = \frac{\overline{x}_2 - \overline{x}_3}{\sqrt{MSE \left(\frac{1}{n_2} + \frac{1}{n_3} \right)}} = \frac{34.0 - 49.25}{\sqrt{138.7 \left(\frac{1}{8} + \frac{1}{8} \right)}} = \frac{-15.25}{5.89} = -2.59$$

We find the p-value using a t-distribution with 21 (error) degrees of freedom, doubling the area below $t = -2.59$ to get p-value $= 2(0.0085) = 0.017$. This is a small p-value so we have evidence of a difference in mean number of ants between peanut butter and ham & pickles sandwiches, with ants seeming to prefer ham & pickles.

8.47 We have three pairs to test. We first test $H_0 : \mu_{DM} = \mu_{LD}$ vs $H_a : \mu_{DM} \neq \mu_{LD}$. The test statistic is

$$t = \frac{\overline{x}_{DM} - \overline{x}_{LD}}{\sqrt{MSE \left(\frac{1}{n_{DM}} + \frac{1}{n_{LD}} \right)}} = \frac{7.859 - 5.987}{\sqrt{6.48 \left(\frac{1}{10} + \frac{1}{9} \right)}} = 1.60.$$

This is a two-tail test so the p-value is twice the area above 1.60 in a t-distribution with $df = 25$. We see that the p-value is $2(0.061) = 0.122$. We don't find convincing evidence for a difference in mean weight gain between the dim light condition and the light/dark condition.

We next test $H_0 : \mu_{DM} = \mu_{LL}$ vs $H_a : \mu_{DM} \neq \mu_{LL}$. The test statistic is

$$t = \frac{\overline{x}_{DM} - \overline{x}_{LL}}{\sqrt{MSE \left(\frac{1}{n_{DM}} + \frac{1}{n_{LL}} \right)}} = \frac{7.859 - 11.010}{\sqrt{6.48 \left(\frac{1}{10} + \frac{1}{9} \right)}} = -2.69.$$

This is a two-tail test so the p-value is twice the area below -2.69 in a t-distribution with $df = 25$. We see that the p-value is $2(0.0063) = 0.0126$. At a 5% level, we do find a difference in mean weight gain between the dim light condition and the bright light condition, with higher mean weight gain in the bright light condition.

Finally, we test $H_0 : \mu_{LD} = \mu_{LL}$ vs $H_a : \mu_{LD} \neq \mu_{LL}$. The test statistic is

$$t = \frac{\overline{x}_{LD} - \overline{x}_{LL}}{\sqrt{MSE \left(\frac{1}{n_{LD}} + \frac{1}{n_{LL}} \right)}} = \frac{5.987 - 11.010}{\sqrt{6.48 \left(\frac{1}{9} + \frac{1}{9} \right)}} = -4.19.$$

This is a two-tail test so the p-value is twice the area below -4.19 in a t-distribution with $df = 25$. We see that the p-value is $2(0.00015) = 0.0003$. There is strong evidence of a difference in mean weight gain between the light/dark condition and the bright light condition, with higher mean weight gain in the bright light condition.

8.49   (a) The p-value for the ANOVA table is essentially zero, so we have strong evidence that there are differences in mean time spent in darkness among the six treatment combinations. Since all the sample sizes are the same, the groups most likely to show a difference in population means are the ones with the sample means farthest apart. These groups are IE:HC (impoverished environment with no added stress, $\overline{x}_1 = 192$) and SE:SD (standard environment with added stress, $\overline{x}_5 = 438$). Likewise, the groups least likely to show a difference in population means are the ones with the sample means closest together. These groups are IE:HC (impoverished environment with no added stress, $\overline{x}_1 = 192$) and SE:HC (standard environment with no added stress, $\overline{x}_2 = 196$).

   (b) Given the six means, there appear to be two distinct groups. The four means (all environments with no added stress, along with the enriched environment with added stress) are all somewhat similar, while the other two means (impoverished or standard environment with added stress) appear to be much larger than the other four and similar to each other. Only the enriched environment with its opportunities for exercise appears to have conferred some immunity to the added stress.

   (c) Let μ_1 and μ_6 be the means for mice in IE:HC and EE:SD conditions, respectively. We test $H_0 : \mu_1 = \mu_6$ vs $H_a : \mu_1 \neq \mu_6$. The test statistic is

$$t = \frac{\overline{x}_1 - \overline{x}_6}{\sqrt{MSE\left(\frac{1}{n_1} + \frac{1}{n_6}\right)}} = \frac{192 - 231}{\sqrt{2469.9\left(\frac{1}{8} + \frac{1}{8}\right)}} = -1.57.$$

   This is a two-tail test so the p-value is twice the area below -1.57 in a t-distribution with $df = 42$. We see that the p-value is $2(0.062) = 0.124$. We do not reject H_0, and do not find convincing evidence of a difference in mean time spent in darkness between mice in the two groups. Prior exercise in an enriched environment may help eliminate the effects of the added stress.

8.51 We have 25 degrees of freedom for the $MSE = 2469.9$ so the t-value for a 95% confidence interval is $t^* = 2.06$. The sample size in each group is 8, so the value of LSD for comparing any two means is

$$LSD = 2.06\sqrt{2469.9\left(\frac{1}{8} + \frac{1}{8}\right)} = 51.2$$

We write the group means down in increasing order

```
Group: IE:HC   SE:HC   EE:HC   EE:SD   IE:SD   SE:SD
Mean:   192     196     205     231     392     438
```

The first four sample means are all within 51.2 of each other so there are no significant differences in mean time in darkness for any of the non-stressed mice groups or the stressed group that has an enriched environment. The mean time is higher for the other two stressed groups (both means are more than 52.1 seconds above the EE:SD group mean), although those two are not significantly different from each other.

Section 9.1 Solutions

9.1 The estimates for β_0 and β_1 are given in two different places in this output, with the output in the table having more digits of accuracy. We see that the estimate for the intercept β_0 is $b_0 = 29.3$ or 29.266 and the estimate for the slope β_1 is $b_0 = 4.30$ or 4.2969. The least squares line is $\widehat{Y} = 29.3 + 4.30X$.

9.3 The estimates for β_0 and β_1 are given in the "Estimate" column of the computer output. We see that the estimate for the intercept β_0 is $b_0 = 77.44$ and the estimate for the slope β_1 is $b_1 = -15.904$. The least squares line is $\widehat{Y} = 77.44 - 15.904 \cdot Score$.

9.5 The slope is $b_1 = -8.1952$. The null and alternative hypotheses for testing the slope are $H_0 : \beta_1 = 0$ vs $H_a : \beta_1 \neq 0$. We see in the output that the p-value is 0.000, so there is strong evidence that the explanatory variable X is an effective predictor of the response variable Y.

9.7 The slope is $b_1 = -0.3560$. The null and alternative hypotheses for testing the slope are $H_0 : \beta_1 = 0$ vs $H_a : \beta_1 \neq 0$. We see in the output that the p-value is 0.087. At a 5% level, we do not find evidence that the explanatory variable *Dose* is an effective predictor of the response variable for this model.

9.9 A confidence interval for the slope β_1 is given by $b_1 \pm t^* \cdot SE$. For a 95% confidence interval with $df = n - 2 = 22$, we have $t^* = 2.07$. We see from the output that $b_1 = -8.1952$ and the standard error for the slope is $SE = 0.9563$. A 95% confidence interval for the slope is

$$
\begin{array}{ccc}
b_1 & \pm & t^* \cdot SE \\
-8.1952 & \pm & 2.07(0.9563) \\
-8.1952 & \pm & 1.980 \\
-10.1752 & \text{to} & -6.2152
\end{array}
$$

We are 95% confident that the population slope β_1 for this model is between -10.18 and -6.22. We can't give a more informative interpretation without some context for the variables.

9.11 We are testing $H_0 : \rho = 0$ vs $H_a : \rho > 0$. To find the test statistic, we use

$$ t = \frac{r\sqrt{n-2}}{\sqrt{1-r^2}} = \frac{0.35\sqrt{28}}{\sqrt{1-(0.35)^2}} = 1.98 $$

This is a one-tail test, so the p-value is the area above 1.98 in a t-distribution with $df = n - 2 = 28$. We see that the p-value is 0.029. At a 5% level, we have sufficient evidence for a positive linear association between the two variables.

9.13 We are testing $H_0 : \rho = 0$ vs $H_a : \rho \neq 0$. To find the test statistic, we use

$$ t = \frac{r\sqrt{n-2}}{\sqrt{1-r^2}} = \frac{0.28\sqrt{98}}{\sqrt{1-(0.28)^2}} = 2.89 $$

This is a two-tail test, so the p-value is twice the area above 2.89 in a t-distribution with $df = n - 2 = 98$. We see that the p-value is $2(0.0024) = 0.0048$. We find strong evidence of a linear association between the two variables.

9.15 (a) The two variables most strongly positively correlated are *Height* and *Weight*. The correlation is $r = 0.619$ and the p-value is 0.000 to three decimal places. A positive correlation in this context means that taller people tend to weigh more.

(b) The two variables most strongly negatively correlated are GPA and $Weight$. The correlation is $r = -0.217$ and the p-value is 0.000 to three decimal places. A negative correlation in this context means that heavier people tend to have lower grade point averages.

(c) At a 5% significance level, almost all pairs of variables have a significant correlation. The only pair not significantly correlated is hours of $Exercise$ and hours of TV with $r = 0.010$ and p-value=0.852, Based on these data, we find no convincing evidence of a linear association between time spent exercising and time spent watching TV.

9.17 (a) On the scatterplot, we have concerns if there is a curved pattern (there isn't) or variability from the line increasing or decreasing in a consistent way (it isn't) or extreme outliers (there aren't any). We do not have any strong concerns about using these data to fit a linear model.

(b) Using the fitted least squares line in the output, for a student with a verbal SAT score of 650, we have

$$\widehat{GPA} = 2.03 + 0.00189 VerbalSAT = 2.03 + 0.00189(650) = 3.2585$$

The model predicts that a student with a 650 on the verbal portion of the SAT exam will have about a 3.26 GPA at this college.

(c) From the computer output the sample slope is $b_1 = 0.00189$. This means that an additional point on the Verbal SAT exam will give a predicted increase in GPA of 0.00189. (And, likewise, a 100 point increase in verbal SAT will raise predicted GPA by 0.189.)

(d) From the computer output the test statistic for testing $H_0 : \beta_1 = 0$ vs $H_a : \beta_1 \neq 0$ is $t = 6.99$ and the p-value is 0.000. This small p-value gives strong evidence that the verbal SAT score is effective as a predictor of grade point average.

(e) In the output we see that R^2 is 12.5%. This tells us that 12.5% of the variability in grade point averages can be explained by verbal SAT scores.

9.19 (a) Because the variable uses z-scores, we look to see if any of the values for the variable $GMdensity$ are larger than 2 or less than -2. We see that there is only one value outside this range, a point with $GMdensity \approx -2.2$ which is less than -2. This participant has a normalized grey matter score more than two standard deviations below the mean and has about 140 Facebook friends.

(b) On the scatterplot, we have concerns if there is a curved pattern (there isn't) or variability from the line increasing or decreasing (it isn't) or extreme outliers (there aren't any). We do not have any serious concerns about using these data to fit a linear model.

(c) In the output the correlation is $r = 0.436$ and the p-value is 0.005. This is a small p-value so we have strong evidence of a linear relationship between number of Facebook friends and grey matter density.

(d) The least squares line is $\widehat{FBfriends} = 367 + 82.4 \cdot GMdensity$. For a person with a normalized grey matter score of 0, the predicted number of Facebook friends is 367. For a person with $GMDensity$ one standard deviation above the mean, the predicted number of Facebook friends is $367 + 82.4(1) = 449.4$. For a person with grey matter density one standard deviation below the mean, the predicted number of Facebook friends is $367 + 82.4(-1) = 284.6$.

(e) The p-value for a test of the slope is 0.005, exactly matching the p-value for the test of correlation. In fact, if we calculate the t-statistic for testing the correlation using $r = 0.436$ and $n = 40$ we have

$$t = \frac{r\sqrt{n-2}}{\sqrt{1-r^2}} = \frac{0.436\sqrt{40-2}}{\sqrt{1-(0.436)^2}} = 2.99$$

which matches the t-statistic for the slope in the computer output.

(f) We see in the computer output that $R^2 = 19.0\%$. This tell us that 19% of the variability in number of Facebook friends can be explained by the normalized grey matter density in the areas of the brain associated with social perception. Since 19% is not very large, many other factors are involved in explaining the other 81% of the variability in number of Facebook friends.

9.21 (a) For a pH reading of 6.0 we have

$$\widehat{AvgMercury} = 1.53 - 0.152 \cdot pH = 1.53 - 0.152(6) = 0.618$$

The model predicts that fish in lakes with a pH of 6.0 will have an average mercury level of 0.618.

(b) The estimated slope is $b_1 = -0.152$. This means that as pH increases by one unit, predicted average mercury level in fish will go down by 0.152 units.

(c) The test statistic is $t = -5.02$, and the p-value is essentially zero. Since this is a very small p-value we have strong evidence that the pH of a lake is effective as a predictor of mercury levels in fish.

(d) The estimated slope is $b_1 = -0.152$ and the standard error is $SE = 0.03031$. For 95% confidence we use a t-distribution with $53 - 2 = 51$ degrees of freedom to find $t^* = 2.01$. The confidence interval for the slope is

$$
\begin{array}{rcl}
b_1 & \pm & t^* \cdot SE \\
-0.152 & \pm & 2.01(0.03031) \\
-0.152 & \pm & 0.0609 \\
-0.2129 & \text{to} & -0.0911
\end{array}
$$

Based on these data we are 95% sure that the slope (increase in mercury for a one unit increase in pH) is somewhere between -0.213 and -0.091.

(e) We see that R^2 is 33.1%. This tells us that 33.1% of the variability in average mercury levels in fish can be explained by the pH of the lake water that the fish come from.

9.23 (a) Since 79% gives the percent of variation accounted for, it is a value of R-squared.

(b) Since precipitation is accounting for prevalence of virus, the response variable is prevalence of the virus and the explanatory variable is precipitation.

(c) Since $R^2 = 0.79$, the correlation is $r = \sqrt{0.79} = 0.889$. The correlation might be either 0.889 or -0.889, but we are told that prevalence increased as precipitation increased, so the correlation is positive. We have $r = 0.889$.

9.25 (a) The cases are countries of the world.

(b) The scatterplot is shown below. We have some mild concern about the extreme $Health$ value on the right with more than 25% of government expenditures on health care. The conditions are mostly met and it is probably acceptable to proceed with the linear regression but with some caution.

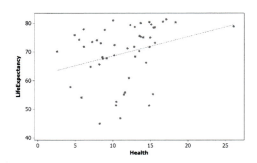

(c) Here is some output for fitting this model.

```
The regression equation is LifeExpectancy = 61.9 + 0.672 Health

Predictor     Coef   SE Coef      T      P
Constant    61.936     3.930  15.76  0.000
Health      0.6717    0.3203   2.10  0.041

S = 9.61686    R-Sq = 8.4%    R-Sq(adj) = 6.5%
```

The slope is $b_1 = 0.672$, so we predict that an increase in 1% expenditure on health care would correspond to an increase in life expectancy of about 0.67 years.

(d) We have $df = 48$ so $t^* = 2.01$. We find a 95% confidence interval by taking $b_1 \pm t^* \cdot SE = 0.672 \pm 2.01(0.320) = (0.029, 1.31)$. We are 95% confident the slope for predicting life expectancy using health expenditure for all countries is between 0.029 and 1.31.

(e) Since our confidence interval does not contain zero we find that the percentage of government expenditure on health is an effective predictor of life expectancy at a 5% level. We can also reach this conclusion by considering the p-value $(0.41 < 0.05)$ given for testing the slope in the regression output.

(f) The population slope for all countries $(\beta_1 = 0.674)$ is very similar to the slope found from this sample $(b_1 = 0.672)$. We also see that $\beta_1 = 0.674$ is easily captured within our confidence interval of 0.029 to 1.31.

(g) In the output we see that $R^2 = 8.4\%$. Only 8.4% of the variability in life expectancy in countries is explained by the percentage of the budget spent on health care.

9.27 (a) Here is a scatterplot of life expectancy vs birthrate for all countries.

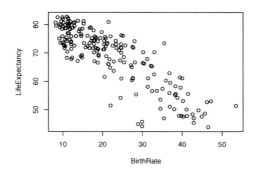

There appears to be a strong negative linear association between birth rate and life expectancy with no obvious curvature and relatively consistent variability.

(b) Yes. We have data on the entire population (all countries) so we can compute $\rho = -0.85$, which is different from 0.

(c) We already have data on (essentially) the entire population so we don't need to use statistical inference to determine where the population correlation might be located.

(d) The slope of the linear regression model is $\beta = -0.8155$, so the predicted life expectancy decreases by about 0.82% for every one percent increase in birth rate of a country.

(e) We cannot conclude that lowering the birthrate in a country would increase its life expectancy. We cannot make conclusions about causality because this is observational data, not data from a randomized experiment.

Section 9.2 Solutions

9.29 The hypotheses are H_0 : The model is ineffective vs H_a : The model is effective. We see in the ANOVA table that the F-statistic is 21.85 and the p-value is 0.000. This is a small p-value so we reject H_0. We conclude that the linear model is effective.

9.31 The hypotheses are H_0 : The model is ineffective vs H_a : The model is effective. We see in the ANOVA table that the F-statistic is 2.18 and the p-value is 0.141. This is a relatively large p-value so we do not reject H_0. We do not find evidence that the linear model is effective.

9.33 We see in the table that the total degrees of freedom is $n - 1 = 175$, so the sample size is 176. To calculate R^2, we use

$$R^2 = \frac{SSModel}{SSTotal} = \frac{3396.8}{30450.5} = 0.112$$

We see that $R^2 = 11.2\%$.

9.35 We see in the table that the total degrees of freedom is $n - 1 = 343$, so the sample size is 344. To calculate R^2, we use

$$R^2 = \frac{SSModel}{SSTotal} = \frac{10.380}{1640.951} = 0.006$$

We see that $R^2 = 0.6\%$.

9.37 The sum of squares for Model and Error add up to Total sum of squares, so

$$SSE = SSTotal - SSModel = 5820 - 800 = 5020$$

The degrees of freedom for the model is 1, since there is one explanatory variable. The sample size is 40 so the Total df is $40 - 1 = 39$ and the Error df is then $40 - 2 = 38$. We calculate the mean squares by dividing sums of squares by degrees of freedom

$$MSModel = \frac{800}{1} = 800 \qquad \text{and} \qquad MSError = \frac{5020}{38} = 132.1$$

The F-statistic is

$$F = \frac{MSModel}{MSError} = \frac{800}{132.1} = 6.06$$

These values are all shown in the following ANOVA table:

Source	d.f.	SS	MS	F-statistic	p-value
Model	1	800	800	6.06	0.0185
Error	38	5020	132.1		
Total	39	5820			

The p-value is found using the upper-tail of an F-distribution with $df = 1$ for the numerator and $df = 38$ for the denominator. For the F-statistic of 6.06, we see that the p-value is 0.0185.

9.39 The sum of squares for Model and Error add up to Total sum of squares, so

$$SSModel = SSTotal - SSE = 23,693 - 15,571 = 8,122$$

The degrees of freedom for the model is 1, since there is one explanatory variable. The sample size is 500 so the Total df is $500 - 1 = 499$ and the Error df is then $500 - 2 = 498$. We calculate the mean squares by dividing sums of squares by degrees of freedom

$$MSModel = \frac{8122}{1} = 8122 \quad \text{and} \quad MSError = \frac{15571}{498} = 31.267$$

The F-statistic is

$$F = \frac{MSModel}{MSError} = \frac{8122}{31.267} = 259.76$$

These values are all shown in the following ANOVA table:

Source	d.f.	SS	MS	F-statistic	p-value
Model	1	8122	8122	259.76	0.000
Error	498	15571	31.267		
Total	499	23693			

The p-value is found using the upper-tail of an F-distribution with $df = 1$ for the numerator and $df = 498$ for the denominator. We don't really need to look this one up, though. The F-statistic of 259.76 is so large that we can predict that the p-value is essentially zero.

9.41 The hypotheses are H_0 : The model is ineffective vs H_a : The model is effective. We see in the ANOVA table that the F-statistic is 7.44 and the p-value is 0.011. This p-value is significant at a 5% level although not at a 1% level. At a 5% level, we conclude that the linear model to predict calories in cereal using the amount of fiber is effective.

9.43 (a) We see that the predicted grade point average for a student with a verbal SAT score of 550 is given by
$$\widehat{GPA} = 2.03 + 0.00189 VerbalSAT = 2.03 + 0.00189(550) = 3.0695$$

We expect an average GPA of about 3.07 for students who get 550 on the verbal portion of the SAT.

 (b) Since the total degrees of freedom is $n - 1 = 344$, the sample size is 345.

 (c) To calculate R^2, we use
$$R^2 = \frac{SSModel}{SSTotal} = \frac{6.8029}{54.5788} = 0.125$$

We see that $R^2 = 12.5\%$, which tells us that 12.5% of the variability in grade point averages can be explained by Verbal SAT score.

 (d) The hypotheses are H_0 : The model is ineffective vs H_a : The model is effective. We see in the ANOVA table that the F-statistic is 48.84 and the p-value is 0.000. This p-value is very small so we reject H_0. There is evidence that the linear model to predict grade point average using Verbal SAT score is effective.

9.45   (a) The regression equation is $\widehat{Points} = 28.8 - 0.0476 PenMins$. The predicted points for a player with 20 penalty minutes is

$$\widehat{Points} = 28.8 - 0.0476(20) = 27.848$$

We expect players with 20 penalty minutes to average about 27.8 points. The predicted points for a player with 150 penalty minutes is

$$\widehat{Points} = 28.8 - 0.0476(150) = 21.66$$

We expect players with 150 penalty minutes to average about 21.7 points.

(b) The slope is -0.0476. If a player has one additional penalty minute, the predicted number of points goes down by 0.0476.

(c) For a test of the slope, the hypotheses are $H_0 : \beta_1 = 0$ vs $H_a : \beta_1 \neq 0$. We see in the output that the t-statistic is -0.62 and the p-value is 0.543. This is a very large p-value so we conclude that number of penalty minutes is not an effective predictor of number of points.

(d) The hypotheses are H_0 : The model is ineffective vs H_a : The model is effective. We see in the ANOVA table that the F-statistic is 0.38 and the p-value is 0.543. This is a very large p-value so we do not reject H_0. We do not find evidence that the linear model based on penalty minutes is effective at predicting number of points for hockey players.

(e) The p-values for the t-test and ANOVA are exactly the same. This will always be the case for a simple regression model.

(f) In the output we see that $R^2 = 1.7\%$, so 1.7% of the variability in number of points scored is explained by the number of penalty minutes. Here again, we see that penalty minutes does not give us very much information about the number of points a player will score.

9.47   (a) To calculate R^2 based on the information in the ANOVA table, we use

$$R^2 = \frac{SSModel}{SSTotal} = \frac{2.0024}{6.0479} = 0.331$$

We see that $R^2 = 33.1\%$, which tells us that 33.1% of the variability in average mercury levels can be explained by the pH of the lake water.

(b) To find the standard deviation of the error term, s_ϵ, we use the value of SSE from the ANOVA table, which is 4.0455, and the sample size of $n = 53$:

$$s_\epsilon = \sqrt{\frac{SSE}{n-2}} = \sqrt{\frac{4.0455}{51}} = 0.282$$

The standard deviation of the error term (labeled as "S= " in the output) is 0.282

(c) To find the standard error of the slope, we need to use $s_\epsilon = 0.282$ from part (b). Also, the explanatory variable is pH so s_x is the standard deviation of the pH values, which is 1.288. Using $n = 53$, we have

$$SE = \frac{s_\epsilon}{s_x \cdot \sqrt{n-1}} = \frac{0.282}{1.288 \cdot \sqrt{52}} = 0.0303$$

The standard error of the slope in the model (shown in the column labeled "SE Coeff") is 0.0303.

9.49 (a) To find the standard deviation of the error term, s_ϵ, we use the value of SSE from the ANOVA table, which is 47.7760, and the sample size of $n = 345$:

$$s_\epsilon = \sqrt{\frac{SSE}{n-2}} = \sqrt{\frac{47.7760}{343}} = 0.373$$

The standard deviation of the error term is $s_\epsilon = 0.373$.

(b) To find the standard error of the slope, we need to use $s_\epsilon = 0.373$ from part (a). Also, the explanatory variable is $VerbalSAT$ so s_x is the standard deviation of the $VerbalSAT$ values, which is 74.29. Using $n = 345$, we have

$$SE = \frac{s_\epsilon}{s_x \cdot \sqrt{n-1}} = \frac{0.373}{74.29 \cdot \sqrt{344}} = 0.00027$$

The standard error of the slope in the model is $SE = 0.00027$.

9.51 (a) We use technology to see that the correlation between $LifeExpectancy$ and $Health$ for the 50 countries in this sample is $r = 0.290$. The t-statistic for testing $H_0 : \rho = 0$ vs $H_a : \rho \neq 0$ is

$$t = \frac{r\sqrt{n-2}}{\sqrt{1-r^2}} = \frac{0.290\sqrt{50-2}}{\sqrt{1-(0.290)^2}} = 2.10$$

Using two-tails and a t-distribution with 48 degrees of freedom, we find $p-value = 2(0.0205) = 0.041$. This is just below a 5% significance level, so we have some evidence of an association between amounts of life expectancy and government health expenditures.

(b) The regression equation is $\widehat{LifeExpectancy} = 61.9 + 0.672Health$. For a test of the slope, $H_0 : \beta_1 = 0$ vs $H_a : \beta_1 \neq 0$ we obtain the output below

```
Predictor    Coef   SE Coef      T       P
Constant   61.936     3.930   15.76   0.000
Health      0.6717    0.3203   2.10   0.041
```

We see that the t-statistic is 2.10 and the p-value is 0.041, so we have evidence, at a 5% significance level, of a relationship between life expectancy and government expenditures for health care.

(c) We use technology to complete an ANOVA test for the effectiveness of this model. The relevant computer output is

```
Source           DF        SS       MS      F      P
Regression        1    406.68   406.68   4.40   0.041
Residual Error   48   4439.23    92.48
Total            49   4845.90
```

We see that the F-statistic is 4.40 with a p-value of 0.041. This gives some evidence for this being an effective model.

(d) This model for predicting life expectancy based on health expenditures is moderately effective. The p-values for both the t-test and ANOVA (which match at 0.041) are just below a 5% significance level.

9.53 (a) Here is some ANOVA output for predicting $Size$ of the California homes using the $Price$. The $SSTotal$ value should be the same for any other predictor of $Size$.

```
Analysis of Variance
Source          DF     SS       MS      F      P
Regression       1   25.362   25.362  21.10  0.000
Residual Error  28   33.658    1.202
Total           29   59.020
```

The sum of squares for the "Total" column is 59.02 which is the total variability (sum of squared differences from the mean) for the sizes of the 30 homes in this sample.

(b) Here are sample correlations with *Size* for each of the potential predictors in the **HomesForSaleCA** dataset.

```
          Price    Beds    Baths
Size      0.656   0.642   0.920
```

If we square these to get the percent of total variability in *Size* that each predictor explains, we get R^2 values of 43.0%, 41.2%, and 84.6% respectively. This indicates that number of bathrooms would be the most effective predictor of the size of these homes.

(c) In part (b) we see that $R^2 = 84.6\%$ is the portion of the variability in size of these homes that is explained by the number of bathrooms. Since $SSTotal = 59.02$ in part (a), the amount explained by the *Baths* variable is $SSModel = 0.846 \cdot SSTotal = 0.846(59.02) = 49.93$. This can be confirmed in the ANOVA output below for fitting a regression model to predict *Size* based on *Baths*.

```
Analysis of Variance
Source          DF     SS       MS       F       P
Regression       1   49.927   49.927  153.74  0.000
Residual Error  28    9.093    0.325
Total           29   59.020
```

(d) We see in part (b) that the weakest predictor of *Size* (smallest correlation) is number of bedrooms ($r = 0.642$). We have $R^2 = 0.642^2 = 0.412$, so 41.2% of the total variability in home sizes is explained by the number of bedrooms. Using the total variability in part (a) we see that the amount explained by *Beds* is $SSModel = 0.412(59.02) = 24.3$.

(e) The output below shows the ANOVA table for assessing the effectiveness as *Beds* as a predictor of *Size*.

```
Analysis of Variance
Source          DF     SS       MS      F      P
Regression       1   24.343   24.343  19.66  0.000
Residual Error  28   34.677    1.238
Total           29   59.020
```

The small p-value (essentially zero) gives strong evidence that, even though it is the weakest of the three potential predictors, *Beds* still gives an effective model for predicting the size of California homes.

Section 9.3 Solutions

9.55   (a) The confidence interval for the mean response is always narrower than the prediction interval for the response, so in this case the confidence interval for the mean response is interval A (10 to 14) and the prediction interval for the response is interval B (4 to 20).

  (b) The predicted value is in the center of both intervals, so we can use the average of the endpoints for either interval to find the predicted value is $(10+14)/2=12$.

9.57   (a) The confidence interval for the mean response is always narrower than the prediction interval for the response, so in this case the confidence interval for the mean response is interval B (4.7 to 5.3) and the prediction interval for the response is interval A (2.9 to 7.1).

  (b) The predicted value is in the center of both intervals, so we can use the average of the endpoints for either interval to find the predicted value is $(4.7+5.3)/2=5$.

9.59   (a) The 95% confidence interval for the mean response is 6.535 to 8.417. We are 95% confident that for mice that eat 50% of calories during the day, the average weight gain will be between 6.535 grams and 8.417 grams.

  (b) The 95% prediction interval for the response is 2.786 to 12.166. We are 95% confident that a mouse that eats 50% of its calories during the day will gain between 2.786 grams and 12.166 grams.

9.61   (a) The 95% confidence interval for the mean response is 122.0 to 142.0. We are 95% confident that the average number of calories for all cereals with 10 grams of sugars per cup will be between 122 and 142 calories per cup.

  (b) The 95% prediction interval for the response is 76.6 to 187.5. We are 95% confident that a cereal with 10 grams of sugar will have between 76.6 and 187.5 calories.

9.63   (a) The size of a home is a significant predictor of the price. The estimated slope from the data is 470.77 and the p-value for testing this coefficient is essentially zero.

  (b) We find a point estimate for the price of a 2,000 square foot home by substituting $Size = 2.0$ (since size is measured in 1,000's of square feet) into the fitted regression equation. This gives

$$\widehat{Price} = -369.63 + 470.77(2) = 571.91$$

The price of a 2,000 square foot home in NY is predicted to be \$571,910.

  (c) We use technology to find a 90% confidence interval for the average price of (445.3, 698.6). We are 90% confident that the average price of 2,000 square foot homes in New York state is between \$445.28 thousand and \$698.44 thousand.

  (d) We use technology to find a 90% prediction interval of (-133.2, 1277.0). We are 90% confident that a 2,000 square foot house in New York would be worth between 0 and \$1.27 million. (The price of a home can't be negative.)

9.65   (a) For a student who gets a 500 on the Verbal SAT exam, the predicted GPA is

$$\widehat{GPA} = 2.03 + 0.00189(500) = 2.98$$

For a student who gets a 700 on the Verbal SAT exam, the predicted GPA is

$$\widehat{GPA} = 2.03 + 0.00189(700) = 3.35$$

(b) For the regression intervals for each Verbal SAT score we have the following

    i. Using technology, we see that a 95% confidence interval for mean GPA for students with a 500 Verbal SAT score is 2.92 to 3.04. We are 95% confident that the mean grade point average for all students at this university who have a 500 Verbal SAT score is between 2.92 and 3.04.

    ii. Using technology, we see that a 95% prediction interval for GPA for students with a 500 Verbal SAT score is 2.24 to 3.72. We are 95% confident that the grade point average for students at this university who have a 500 Verbal SAT score is between 2.24 and 3.72.

    iii. Using technology, we see that a 95% confidence interval for mean GPA for students with a 700 Verbal SAT score is 3.29 to 3.43. We are 95% confident that the mean grade point average for all students at this university who have a 700 Verbal SAT score is between 3.29 and 3.43.

    iv. Using technology, we see that a 95% prediction interval for GPA for students with a 700 Verbal SAT score is 2.62 to 4.10. We are 95% confident that the grade point average for students at this university who have a 700 Verbal SAT score is between 2.62 and 4.00. (At this university, the highest possible GPA is 4.0.)

9.67 To calculate the 95% confidence intervals and prediction intervals for final scores we use a t-distribution with $20 - 2 = 18$ degrees of freedom to find $t^* = 2.10$. We see from the computer output that the standard deviation of the error is $s_\epsilon = 3.598$ and from the summary statistics for the predictor ($First$) that $\bar{x} = -0.550$ and $s_x = 3.154$. We use these values to find the requested intervals.

(a) When the first round score is 0, we see that the predicted final round score is $\widehat{Final} = 0.1617 + 01.4756(0) = 0.1617$. To find a 95% confidence interval for the mean final score after a 0 on the first round, we have

$$\widehat{Final} \quad \pm \quad t^* s_\epsilon \sqrt{\frac{1}{n} + \frac{(x^* - \bar{x})^2}{(n-1)s_x^2}}$$

$$0.1617 \quad \pm \quad 2.10(3.598)\sqrt{\frac{1}{20} + \frac{(0 - (-0.550))^2}{19(3.154^2)}}$$

$$0.1617 \quad \pm \quad 1.716$$

$$-1.55 \quad \text{to} \quad 1.88$$

We are 95% sure that the mean final score of golfers who shoot a 0 on the first round at the Masters is between -1.55 and 1.88.

(b) When the first round score is -5, we see that the predicted final round score is $\widehat{Final} = 0.1617 + 1.4756(-5) = -7.216$. To find a 95% prediction interval for the final score after a -5 on the first round, we have

$$\widehat{Final} \quad \pm \quad t^* s_\epsilon \sqrt{1 + \frac{1}{n} + \frac{(x^* - \bar{x})^2}{(n-1)s_x^2}}$$

$$-7.216 \quad \pm \quad 2.10(3.598)\sqrt{1 + \frac{1}{20} + \frac{(-5 - (-0.550))^2}{19(3.154^2)}}$$

$$-7.216 \quad \pm \quad 8.119$$

$$-15.34 \quad \text{to} \quad 0.90$$

We are 95% sure that the four round score of a golfer who shoots a -5 on the first round at the Masters is between -15.34 and 0.90. Since golf scores are only integers we could report this prediction interval as -15 to $+1$ for the golfer's final score.

(c) When the first round score is $+3$, we see that the predicted final round score is $\widehat{Final} = 0.1617 + 1.4756(3) = 4.589$. To find a 95% confidence interval for the mean final score after a $+3$ on the first round, we have

$$\widehat{Final} \quad \pm \quad t^* s_\epsilon \sqrt{\frac{1}{n} + \frac{(x^* - \overline{x})^2}{(n-1)s_x^2}}$$

$$4.589 \quad \pm \quad 2.10(3.598)\sqrt{\frac{1}{20} + \frac{(3 - (-0.550))^2}{19(3.154^2)}}$$

$$4.589 \quad \pm \quad 2.581$$

$$2.01 \quad \text{to} \quad 7.17$$

We are 95% sure that the mean final score of golfers who shoot a $+3$ on the first round at the Masters is between 2.01 and 7.17.

Section 10.1 Solutions

10.1 There are four explanatory variables: $X1$, $X2$, $X3$, and $X4$. The one response variable is Y.

10.3 The predicted response is

$\widehat{Y} = 43.4 - 6.82X1 + 1.70X2 + 1.70X3 + 0.442X4 = 43.4 - 6.82(5) + 1.70(7) + 1.70(5) + 0.442(75) = 62.85.$

The residual is $60 - 62.85 = -2.85$.

10.5 The coefficient of $X1$ is -6.820 and the p-value for testing this coefficient is 0.001.

10.7 Looking at the p-values, we see that variable $X1$ is the only one that is significant at a 1% level.

10.9 Looking at the p-values, we see that $X2$ is least significant.

10.11 We see that $R^2 = 99.8\%$ so 99.8% of the variability in Y is explained by the model.

10.13 The predicted response is

$$\begin{aligned} \widehat{Y} &= -61 + 4.71X1 - 0.25X2 + 6.46X3 + 1.50X4 - 1.32X5 \\ &= -61 + 4.71(15) - 0.25(40) + 6.46(10) + 1.50(50) - 1.32(95) \\ &= 13.85. \end{aligned}$$

The residual is $Y - \widehat{Y} = 20 - 13.85 = 6.15$.

10.15 The coefficient or slope of $X1$ is 4.715 and the p-value for testing this coefficient is 0.053.

10.17 Looking at the p-values, we see that variables $X3$ and $X4$ are significant at a 5% level and variables $X1$, $X2$, and $X5$ are not.

10.19 Looking at the p-values, we see that $X4$ is most significant.

10.21 Yes, the p-value from the ANOVA table is 0.000 so the model is effective.

10.23   (a) The predicted price for a 2500 square foot home with 4 bedrooms and 2.5 baths is

$$\begin{aligned} \widehat{Price} &= -217 + 331SizeSqFt - 135Beds + 200Baths \\ &= -217 + 0.331(2500) - 135(4) + 200(2.5) \\ &= 570.5. \end{aligned}$$

The predicted price for this house is \$570,500.

(b) The largest coefficient is 200, the coefficient for the number of bathrooms.

(c) The most significant predictor in this model is $SizeSqFt$ with $t = 4.55$ and p-value $= 0.000$ – even though the size of its coefficient, $\hat{\beta}_1 = 0.331$, is much smaller than the coefficients of the other two predictors.

(d) All three predictors are significant at the 5% level.

(e) If all else stays the same (such as number of bedrooms and number of bathrooms), a house with 1 more square foot in area is predicted to cost 0.331 thousand dollars (\$331) more.

(f) The model is effective at predicting price, in the sense that at least one of the explanatory variables are useful, since the p-value is 0.000.

(g) We see that 46.7% of the variability in prices of new homes can be explained by the area in square feet, the number of bedrooms, and the number of bathrooms.

10.25 (a) The last column in the table above represents the p-value. We see that both Internet and BirthRate have p-values less then 0.05, so in this model percentage of the population with internet and the countries birth rate are significant predictors of life expectancy. The p-value for birth rate is much smaller, so this is the most significant predictor.

(b) If we plug these values into our equation we get $\widehat{Life} = 77.30 + 0.133(20) - 0.017(2.5) + 0.112(75) - 0.555(30) = 71.67$. So we predict the life expectancy for this country would be 71.67.

(c) Since the coefficient of internet is positive, the predicted life expectancy would increase.

10.27 The degrees of freedom for the variability explained by the regression model, 3, equals the number of predictors in the model.

10.29 $R^2 = \dfrac{SSModel}{SSTotal} = \dfrac{4327.7}{9999.1} = 0.433$. Although we don't know the specific predictors, 43.3% of the variability in the horse prices is explained by the three predictors.

10.31 (a) The sentences quoted are talking about the percent of variation accounted for, so the quantity being discussed is the value of R-squared.

(b) We expect the p-value to be very small. With such a large R^2 – 98% of the variability accounted for – this model is clearly very effective at predicting prevalence of hantavirus in mice.

10.33 Here is some typical computer output for this model:

```
The regression equation is
BM Gain = - 1.95 + 0.00096 Corticosterone + 0.116 DayPct + 0.904 Consumption
            - 0.000102 Activity


Predictor                Coef     SE Coef       T      P
Constant               -1.953       2.772   -0.70  0.488
Corticosterone       0.000963    0.009201    0.10  0.918
DayPct                0.11614     0.02492    4.66  0.000
Consumption            0.9040      0.5721    1.58  0.128
Activity            -0.0001020   0.0002326  -0.44  0.665

S = 2.25158   R-Sq = 59.4%   R-Sq(adj) = 52.1%


Analysis of Variance
Source          DF       SS       MS      F      P
Regression       4  163.391   40.848   8.06  0.000
Residual Error  22  111.531    5.070
Total           26  274.922
```

(a) The coefficient is 0.116, so if all the other explanatory variables in the model are held constant, the predicted body mass gain goes up 0.116 grams for one more percent of food eaten during the day.

(b) The coefficient is 0.904, so if all the other explanatory variables in the model are held constant, the predicted body mass gain goes up 0.904 grams for one additional gram of food in daily consumption.

(c) The percent of food eaten during the day, *DayPct*, is the most significant variable.

(d) Stress levels, *Corticosterone*, is the least significant variable in this model.

(e) Yes, the model is effective, in the sense that at least one of the explanatory variables are useful for predicting body mass gain, since the p-value from the ANOVA table is 0.000.

(f) We see that 59.4% of the variability in body mass gain can be explained by these four variables.

10.35 (a) Here is some output for predicting *Price* based on *Age* alone. The value for the test of slope is essentially zero, providing strong evidence that *Age* is an effective predictor of Mustang prices in this model.

```
Predictor      Coef  SE Coef      T       P
Constant     30.264    3.440   8.80   0.000
Age          -1.7168   0.3648  -4.71   0.000
```

(b) Here is some output for the model using both *Age* and *Miles* as predictors of *Price*. The p-value for the t-test of the coefficient of *Age* is 0.769 which is not small at all. This does not provide enough evidence to conclude that *Age* is a useful predictor of *Price* in this model.

```
Predictor      Coef  SE Coef      T       P
Constant     30.867    2.787  11.07   0.000
Miles       -0.20495  0.05650  -3.63   0.001
Age          -0.1551   0.5219  -0.30   0.769
```

(c) The discrepancy between parts (a) and (b) occurs because *Age* and *Miles* driven are very strongly related ($r = 0.825$ for these 25 cars). If *Miles* is already in the model we don't need *Age* as well because it has little additional information to predict price that *Miles* doesn't already supply. If *Miles* is not in the model, as in part (a), then *Age* is a useful predictor.

10.37 Here are ANOVA tables for two randomizations of the *Price* values, using *PhotoTime* and *CostColor* as predictors.

```
Source          DF      SS      MS      F      P
Regression       2   29821   14910   2.38   0.122
Residual Error  17  106416    6260
Total           19  136237

Source          DF      SS      MS      F      P
Regression       2    2428    1214   0.15   0.858
Residual Error  17  133808    7871
Total           19  136237
```

These give randomization F-statistics 2.38 and 0.15, respectively. We repeat this to get a total of 1000 randomization F-statistics as shown in the dotplot below.

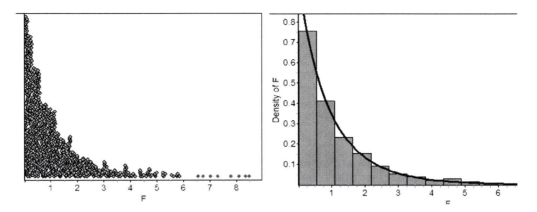

For this simulation we find just 8 of the 1000 randomizations produce an F-statistic as large as the F=6.30 observed in the original sample. This gives an approximate p-value of 8/1000 = 0.008 which agrees nicely with the p-value=0.009 from the F-distribution. The histogram of the randomization distribution above on the right also shows the density for an F-distribution with 2 numerator and 17 denominator degrees of freedom.

Section 10.2 Solutions

10.39 We see that the residuals are pretty evenly distributed around the zero line, without obvious curvature or outliers. This matches the nice linear relationship we see in graph (b).

10.41 We see that there is a large outlier in the scatterplot. Since the outlier is below the line, the residual is negative. Also, the outlier matches a predicted value of about 55, which matches what we see in graph (d).

10.43 The conditions are not all met. Normality of the residuals is acceptable since we see no strong skewness or outliers in the histogram of the residuals. However, we see in the scatterplot of the data with the least squares line that the trend in the data is not really linear. This departure from linearity is also apparent in the scatterplot of residuals against predicted values where we see obvious curvature.

10.45   (a) The arrow is shown on the figure.

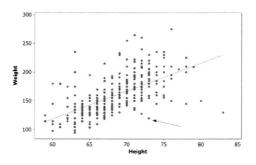

   (b) The predicted weight is $\widehat{Weight} = -170 + 4.82 Height = -170 + 4.82(73) = 181.86$ lbs. The residual is $120 - 181.86 = -61.86$.

   (c) Arrows are shown on both the figures.

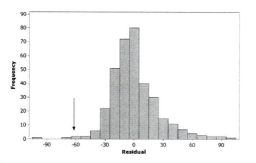 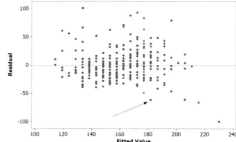

   (d) Looking at the three graphs, the conditions appear to all be met.

10.47 The three relevant plots, a scatterplot with least squares line, a histogram of the residuals, and a residuals versus fitted values plot are shown below.

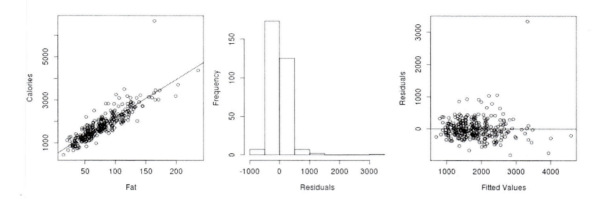

The conditions do not appear to be well met. In particular, there is one very high outlier (someone who eats 6662.2 calories a day!). This outlier is obvious and extreme on the scatterplot with regression line, the histogram of residuals, and the scatterplot of residuals vs fits.

10.49 The three relevant plots, a scatterplot with least squares line, a histogram of the residuals, and a residuals versus fitted values plot are shown below.

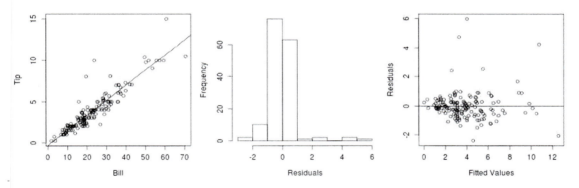

Some of the conditions met, but the very high outliers in the histogram indicate that the residuals are probably not normally distributed. Other than that, there appears to be a relatively strong linear trend in the scatterplot with regression line with reasonably equal variability. There are no big concerns from the scatterplot of residuals vs fits, which seems to show no obvious trends or curvature and relatively equal variability.

10.51 (a) Here is some output for fitting the model with the St. Louis data

```
             Estimate Std. Error t value Pr(>|t|)
(Intercept)  6.40819    0.58764   10.90   <2e-16 ***
Distance     1.09931    0.03307   33.24   <2e-16 ***
```

The prediction equation for St. Louis is $\widehat{Time} = 6.41 + 1.0999 \cdot Distance$. This is similar to the model for Atlanta, with a slightly smaller intercept and slope.

(b) For $Distance = 20$ the expected commute time for the St. Louis model is $\widehat{Time} = 6.41 + 1.099 \cdot 20 = 28.39$ minutes (compared to 31.34 minutes for the same distance commute in Atlanta).

(c) A scatterplot of *Time* vs *Distance*, histogram of the residuals, and residual vs fits plot for the St. Louis data are shown below.

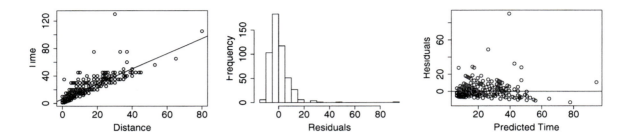

These plots are similar to the corresponding plots for the Atlanta data. In the scatterplot of *Time* vs *Distance*, there is a steady upward trend with longer commutes tending to take longer times. However, there is a very regular boundary along the bottom edge of the graph (possibly determined by the speed limit?) that keeps points fairly close and in the direction of the regression line. The deviations above the line are more scattered with several unusually large positive residuals.

(d) The histogram of the residuals shows a clear skew to the right with a long tail and several large outliers in that direction. The normality condition is not appropriate for these residuals.

(e) The plot of residual vs fits for this model shows a distinct pattern, especially for the negative residuals, of increasing variability of the residuals as the predicted commute time increases. The equal variability condition is not met for this model and sample. Also, the plot shows several very large positive residuals where the commute time is drastically underestimated.

10.53 A histogram of the residuals and plot of residuals versus fitted values are shown below. The histogram is somewhat skewed with some potential large outliers, so we have some minor concerns about the normality condition. The residual plot presents more of a concern. The majority of points indicate a clear downward trend, with the exception of a few extreme outliers. This raises concern about using this linear model.

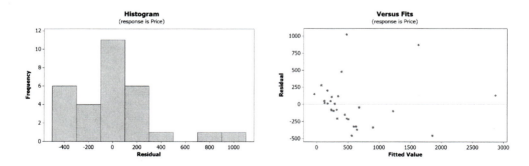

10.55 A histogram of the residuals and plot of residuals versus fitted values are shown below. The histogram is nicely symmetric with no extreme outliers, although the tails don't taper much. The normality condition is probably quite reasonable. The residuals are randomly scattered in a band on either side of the zero line in the residual vs fits plot. We see no concerns with the constant variability condition.

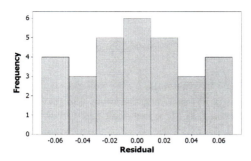 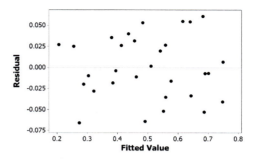

Section 10.3 Solutions

10.57 (a) We should try eliminating the variable $X2$ since it has the largest p-value and hence is the most insignificant in this model.

(b) We see that $R^2 = 15\%$. Eliminating any variable (even insignificant ones) will cause R^2 to decrease. A very small decrease in R^2 would indicate that removing $X2$ was a good idea, whereas a large decrease in R^2 would indicate that removing $X2$ may have been a bad idea, because a larger R^2 generally means a better model.

(c) The p-value is 0.322. Removing an insignificant variable will most likely improve the model, and so cause the ANOVA p-value to decrease. If the p-value decreases it would indicate that removing the insignificant variable was a good idea, if the p-value increases it would indicate that removing the insignificant variable was probably not a good idea.

10.59 Using all five variables, we get the following information:

```
The regression equation is
Profitability = 5.42 + 0.0763 RottenTomatoes - 0.0958 AudienceScore
               - 0.000566 TheatersOpenWeek - 0.000036 BOAverageOpenWeek
               + 0.0278 DomesticGross
```

Predictor	Coef	SE Coef	T	P
Constant	5.420	4.209	1.29	0.200
RottenTomatoes	0.07630	0.04342	1.76	0.082
AudienceScore	-0.09577	0.07334	-1.31	0.194
TheatersOpenWeek	-0.0005657	0.0009960	-0.57	0.571
BOAverageOpenWeek	-0.00003603	0.00007769	-0.46	0.644
DomesticGross	0.02781	0.01577	1.76	0.081

```
S = 6.88934   R-Sq = 6.4%   R-Sq(adj) = 2.3%
Analysis of Variance
```

Source	DF	SS	MS	F	P
Regression	5	365.76	73.15	1.54	0.183
Residual Error	112	5315.85	47.46		
Total	117	5681.61			

Notice that the p-value from ANOVA is 0.183 and there are several insignificant variables. We can definitely make this model better. On the other hand, we see that R^2 is only 6.4% so this is never going to be an outstanding model! To improve the model, answers will vary but it makes sense to eliminate the least significant variable, $BOAverageOpenWeek$. We eliminate only one at a time, even though there are multiple insignificant variables, because it is plausible that the variables are correlated with each other. When we eliminate $BOAverageOpenWeek$, we get the following model:

```
The regression equation is
Profitability = 4.65 + 0.0728 RottenTomatoes - 0.0896 AudienceScore
               - 0.000363 TheatersOpenWeek + 0.0236 DomesticGross
```

Predictor	Coef	SE Coef	T	P
Constant	4.652	3.855	1.21	0.230

RottenTomatoes	0.07279	0.04260	1.71	0.090
AudienceScore	-0.08962	0.07188	-1.25	0.215
TheatersOpenWeek	-0.0003628	0.0008916	-0.41	0.685
DomesticGross	0.02356	0.01280	1.84	0.068

S = 6.86537   R-Sq = 6.3%   R-Sq(adj) = 2.9%

Analysis of Variance

Source	DF	SS	MS	F	P
Regression	4	355.55	88.89	1.89	0.118
Residual Error	113	5326.06	47.13		
Total	117	5681.61			

We see that R^2 has decreased (as expected), but only from 6.4% to 6.3%. Adjusted R^2 has increased from 2.3% to 2.9%, the ANOVA p-value has decreased from 0.183 to 0.118, and s_ϵ decreased from 6.889 to 6.865. All of these facts indicate that removing $BOAverageOpenWeek$ improved the model. We next eliminate the most insignificant remaining variable, $TheatersOpenWeek$, with a p-value of 0.685, and get the following model:

The regression equation is
Profitability = 2.94 + 0.0620 RottenTomatoes - 0.0685 AudienceScore
                + 0.0214 DomesticGross

Predictor	Coef	SE Coef	T	P
Constant	2.935	2.348	1.25	0.214
RottenTomatoes	0.06198	0.03666	1.69	0.093
AudienceScore	-0.06851	0.05993	-1.14	0.255
DomesticGross	0.021354	0.008659	2.47	0.015

S = 6.50362   R-Sq = 6.8%   R-Sq(adj) = 4.6%

Analysis of Variance

Source	DF	SS	MS	F	P
Regression	3	394.03	131.34	3.11	0.029
Residual Error	128	5414.02	42.30		
Total	131	5808.05			

The model has definitely improved and is now significant at the 5% level, since the ANOVA p-value is now 0.029. Notice that the value of R^2 went up, which can only happen when there are missing values (as is the case here.) Normally, R^2 will always stay the same or go down when eliminating variables. Notice also that the value of s_ϵ went from 6.865 to 6.504, giving further evidence that the model has improved. There is one variable, $AudienceScore$, that is not even significant at the 10% level, so we try eliminating that to see if the model improves. We have:

The regression equation is
Profitability = 0.69 + 0.0281 RottenTomatoes + 0.0185 DomesticGross

Predictor	Coef	SE Coef	T	P
Constant	0.694	1.293	0.54	0.592
RottenTomatoes	0.02809	0.02158	1.30	0.195

```
DomesticGross    0.018478    0.008296    2.23    0.028

S = 6.51134    R-Sq = 5.8%    R-Sq(adj) = 4.4%

Analysis of Variance
Source           DF        SS        MS       F       P
Regression        2    338.76    169.38    4.00    0.021
Residual Error  129   5469.29     42.40
Total           131   5808.05
```

The ANOVA p-value went down a bit, but the standard deviation of the error went up. We see that R^2 went down, as we expect. You can decide which of the two previous models you prefer, or if you want to try another one by eliminating the variable *RottenTomatoes* to see what happens.

10.61 (a) A correlation matrix for all the variables is shown. We see that the variables with strong correlations with *BetaPlasma* are *Fiber* and *BetaDiet*. To a smaller extent, but still possibly relevant, are *Age* and *Fat*. It appears that *Alcohol* consumption is not significantly correlated with beta-carotene levels in the body.

```
              BetaPlasma       Age        Fat       Fiber     Alcohol
Age             0.101
                0.073

Fat            -0.092        -0.169
                0.104         0.003

Fiber           0.236         0.045      0.276
                0.000         0.428      0.000

Alcohol        -0.022         0.052      0.186      -0.020
                0.695         0.362      0.001       0.722

BetaDiet        0.225         0.072      0.143       0.483      0.039
                0.000         0.203      0.011       0.000      0.486

Cell Contents: Pearson correlation
               P-Value
```

(b) Let's start with a model with all five explanatory variables, just to see what we get.

```
The regression equation is
BetaPlasma = 92.4 + 0.677 Age - 0.874 Fat + 7.19 Fiber + 0.053 Alcohol
             + 0.0177 BetaDiet

Predictor      Coef    SE Coef       T       P
Constant      92.42      47.92    1.93    0.055
Age          0.6768     0.6943    0.97    0.330
Fat         -0.8739     0.3164   -2.76    0.006
Fiber         7.186      2.191    3.28    0.001
```

```
Alcohol       0.0533     0.8219    0.06   0.948
BetaDiet      0.017745   0.007668  2.31   0.021


S = 174.845   R-Sq = 10.2%   R-Sq(adj) = 8.7%

Analysis of Variance
Source          DF         SS       MS      F      P
Regression       5    1069284   213857   7.00  0.000
Residual Error 309    9446354    30571
Total          314   10515638
```

The ANOVA p-value is 0, but R^2 is only 10.2%. As expected from the correlation analysis, the p-value for *Alcohol* is very large. Let's look at a model with just the other four variables.

```
The regression equation is
BetaPlasma = 92.2 + 0.681 Age - 0.870 Fat + 7.17 Fiber + 0.0178 BetaDiet

Predictor      Coef    SE Coef      T      P
Constant      92.17      47.69   1.93   0.054
Age          0.6809     0.6903   0.99   0.325
Fat         -0.8695     0.3087  -2.82   0.005
Fiber         7.172      2.177   3.29   0.001
BetaDiet   0.017770   0.007646   2.32   0.021


S = 174.564   R-Sq = 10.2%   R-Sq(adj) = 9.0%

Analysis of Variance
Source          DF         SS       MS      F      P
Regression       4    1069155   267289   8.77  0.000
Residual Error 310    9446483    30473
Total          314   10515638
```

We see that R^2 did not go down at all, and the F-statistic got better (increased), adjusted R^2 got better (increased), and s_ϵ got better (decreased). This is a better model than the first one. However, the variable *Age* is not significant in the model, so we try again without that one.

```
The regression equation is
BetaPlasma = 128 - 0.927 Fat + 7.30 Fiber + 0.0182 BetaDiet

Predictor      Coef    SE Coef      T      P
Constant     128.20      30.67   4.18   0.000
Fat         -0.9275     0.3030  -3.06   0.002
Fiber         7.296      2.173   3.36   0.001
BetaDiet   0.018228   0.007632   2.39   0.018


S = 174.556   R-Sq = 9.9%   R-Sq(adj) = 9.0%

Analysis of Variance
```

```
Source             DF        SS       MS      F      P
Regression          3    1039508   346503   11.37  0.000
Residual Error    311    9476130    30470
Total             314   10515638
```

Few quantities except the the F-statistic changed, and the F-statistic continued to go up. All three variables are significant at a 5% level (and almost at a 1% level) so we'll call this our final model. Other answers are also possible.

10.63 (a) We regress *LifeExpectancy* on *Electricity* and obtain the following output:

```
Coefficients:
                Estimate Std. Error t value Pr(>|t|)
(Intercept) 6.767e+01  8.473e-01   79.869  < 2e-16 ***
Electricity 7.567e-04  1.190e-04    6.359 3.06e-09 ***

Residual standard error: 8.023 on 132 degrees of freedom
  (79 observations deleted due to missingness)
Multiple R-squared: 0.2345,    Adjusted R-squared: 0.2287
F-statistic: 40.43 on 1 and 132 DF,  p-value: 3.064e-09
```

The p-value of 3.064×10^{-9} (essentially zero) indicates that *Electricity* is a very significant predictor of *LifeExpectancy*.

(b) GDP is associated with both electricity use ($r = 0.672$) and with life expectancy ($r = 0.578$), so is a potential confounding variable.

(c) We regress *LifeExpectancy* on both *Electricity* and *GDP* and obtain the following output:

```
Coefficients:
                Estimate Std. Error t value Pr(>|t|)
(Intercept) 6.611e+01  8.804e-01   75.089  < 2e-16 ***
Electricity 1.945e-04  1.608e-04    1.210    0.229
GDP         2.551e-04  5.013e-05    5.089 1.39e-06 ***

Residual standard error: 7.58 on 117 degrees of freedom
  (93 observations deleted due to missingness)
Multiple R-squared: 0.3571,    Adjusted R-squared: 0.3462
F-statistic:  32.5 on 2 and 117 DF,  p-value: 5.954e-12
```

The p-value of 0.229 for *Electricity* indicates that after accounting for the information in *GDP*, *Electricity* is no longer a significant predictor of *LifeExpectancy* in this model.

10.65 (a) For the model with size in square feet, the coefficient of $SizeSqFt$ is 0.33058. When we code *Size* in thousands of square feet, we multiply this coefficient by 1000 to get the coefficient of 330.58 in the revised model. Note that predictions will be identical for the same size home with either model. The standard error for the coefficient is also 1000 times larger in the revised model (going from $SE = 0.07262$ to $SE = 72.62$) so the t-statistic ($t = 4.552$) and p-value remain the same.

(b) Given the size of the house and the number of bathrooms, each additional bedroom decreases the predicted price by \$134,000.

(c) The predictors are all highly correlated with each other, making any individual coefficient very difficult to interpret. Number of bedrooms is too highly associated with size of the house and number of bathrooms to draw any meaningful conclusions from the negative coefficient. Also, this is observational data, so we cannot draw any causal conclusions.

10.67 The predicted distance is higher for the carbon bike, by 0.396 miles (as seen by the coefficient of *BikeSteel*. The p-value of 4.74×10^{-9} (essentially zero) indicates that this difference is very significant.

10.69 (a) Without accounting for distance, the steel bike gives a lower predicted commute time. However, in this experiment the commutes ridden on the steel bike were of slightly shorter distance. After accounting for distance, for rides of the same distance the carbon bike gives a lower predicted commute time.

(b) Because the coefficient of *BikeSteel* is positive with *Distance* included in the model, the predicted commute time for a ride of a given distance is longer for the steel bike. Therefore, the steel bike has a lower average speed, and so the coefficient for *BikeSteel* would be negative.

Unit D: Essential Synthesis Solutions

D.1 This is a chi-square goodness-of-fit test. The null hypothesis is that the bills are equally distributed among the three servers while the alternative hypothesis is that they are not equally spread out. In symbols the hypotheses are

$H_0:$ $p_A = p_B = p_C = 1/3$
$H_a:$ Some $p_i \neq 1/3$

The expected count for each cell is $n \cdot p_i = 157 \cdot (1/3) = 52.33$. We compute the chi-square statistic:

$$\chi^2 = \frac{(60 - 52.33)^2}{52.33} + \frac{(65 - 52.33)^2}{52.33} + \frac{(32 - 52.33)^2}{52.33} = 1.124 + 3.068 + 7.898 = 12.09$$

The upper-tail p-value from a chi-square distribution with $df = 2$ is 0.002. This is a small p-value so we reject H_0 and find evidence that the bills are not equally distributed between the three servers. Server C appears to have substantially fewer bills than expected if they were equally distributed, and the result is significant enough to generalize (assuming the sample data are representative of all bills).

D.3 This is a chi-square test on a two-way table. The null hypothesis is that use of a credit card does not differ based on the server and the alternative hypothesis is that there is an association between card use and server.

The expected count for the (Cash, Server A) cell is

$$(60 \cdot 106)/157 = 40.51$$

For the (Cash, Server A) cell, we then compute the contribution to the chi-square statistic as

$$(39 - 40.51)^2/40.51 = 0.056$$

Finding the other expected counts and contributions similarly (or using technology) produces a table with the observed counts in each cell, expected counts below them, and the contributions to the chi-square statistic below that, as in the computer output below.

```
            A       B       C
Cash       39      50      17
          40.51   43.89   21.61
          0.0563  0.8520  0.9816

Card       21      15      15
          19.49   21.11   10.39
          0.1169  1.7708  2.0401

Cell Contents:      Count
                    Expected count
                    Contribution to Chi-square
```

`Pearson Chi-Square = 5.818, DF = 2, P-Value = 0.055`

Adding up all the contributions to the chi-square statistic, we obtain $\chi^2 = 5.818$ (also seen in the computer output). Using the upper tail of a chi-square distribution with $df = 2$, we obtain a p-value of 0.055. This is a borderline p-value, but, at a 5% level, does not provide enough evidence of an association between server and the use of cash or credit.

D.5 This is an analysis of variance test for a difference in means. The sample sizes of the three groups (servers) are all greater than 30, so the normality condition is met and we see from the standard deviations that the condition of relatively equal variability is also met. We proceed with the ANOVA test.

The null hypothesis is that the means are all the same (no association between tip percent and server) and the alternative hypothesis is that there is a difference in mean tip percent between the servers. In symbols, where μ represents the mean tip percentage, the hypotheses are

$$H_0: \quad \mu_A = \mu_B = \mu_C$$
$$H_a: \quad \text{Some } \mu_i \neq \mu_j$$

Using technology with the data in **RestaurantTips** we obtain the analysis of variance table shown below.

```
One-way ANOVA: PctTip versus Server

Source   DF     SS     MS    F      P
Server    2    83.1   41.6  2.19  0.115
Error   154  2917.9   18.9
Total   156  3001.1
```

We see that the F-statistic is 2.19 and the p-value for an F-distribution with 2 and 154 degrees of freedom is 0.115. This is not a small p-value so we do not reject H_0. We do not find convincing evidence of a difference in mean percentage tip between the three servers.

D.7 (a) There appears to be a positive association in the data with larger parties tending to give larger tips. There is one outlier: The (generous) party of one person who left a \$15 tip.

(b) We are testing $H_0: \rho = 0$ vs $H_a: \rho > 0$ where ρ is the correlation between number of guests and size of the tip. As always, the null hypothesis is that there is no relationship while the alternative hypothesis is that there is a relationship between the two variables. The test statistic is

$$t = \frac{r \cdot \sqrt{n-2}}{\sqrt{1-r^2}} = \frac{0.504 \cdot \sqrt{155}}{\sqrt{1-(0.504^2)}} = 7.265$$

This is a one-tailed test, so the p-value is the area above 7.265 in a t-distribution with $df = n - 2 = 155$. We see that the p-value is essentially zero. There is strong evidence of a significant positive linear relationship between these two variables. The tip does tend to be larger if there are more guests.

(c) No, we cannot assume causation since these data do not come from an experiment. An obvious confounding variable is the size of the bill, since the bill tends to be higher for more guests and higher bills generally tend to correspond to higher tips.

D.9 (a) We see that $R^2 = 83.7\%$, which tells us that 83.7% of the variability in the amount of the tip is explained by the size of the bill.

(b) From the output the F-statistic is 797.87 and the p-value is 0.000. There is very strong evidence that this regression line $\widehat{Tip} = -0.292 + 0.182 \cdot Bill$ is effective at predicting the size of the tip.

Unit D: Review Exercise Solutions

D.11 The area above $t = 1.36$ gives a p-value of 0.0970, which is not significant at a 5% level.

D.13 For 5 groups we have $5 - 1 = 4$ degrees of freedom for the chi-square goodness-of-fit statistic. The area above $\chi^2 = 4.18$ gives a p-value of 0.382, which is not significant at a 5% level.

D.15 For a difference in means ANOVA with $k = 6$ groups and and an overall sample size of $n = 100$ we use an F-distribution with $k - 1 = 6 - 1 = 5$ numerator df and $n - k = 100 - 6 = 94$ denominator df. The area above $F = 2.51$ for this distribution gives a p-value of 0.035, which is significant at a 5% level.

D.17 For a multiple regression model with $k = 3$ predictors we have $n - k - 1 = 26 - 3 - 1 = 22$ degrees of freedom for the t-distribution to test an individual coefficient. Accounting for two tails and the area beyond $t = 1.83$ gives a p-value of $2(0.0404) = 0.0808$, which is not significant at a 5% level.

D.19 White blood cell count is a quantitative variable so this is a test for a difference in means between the three groups, which is analysis of variance for difference in means.

D.21 Both of the relevant variables are quantitative, and we can use a test for correlation, a test for slope, or ANOVA for regression.

D.23 The data are counts from the sample within different racial categories. To compare these with the known racial proportions for the entire city, we use a chi-square goodness-of-fit test with the census proportions as the null hypothesis.

D.25 They are using many quantitative variables to develop a multiple regression model. To test its effectiveness, we use ANOVA for regression.

D.27 Whether or not a case gets settled out of court is categorical, and the county is the other categorical variable, so this is a test between two categorical variables. A chi-square test for association is most appropriate.

D.29 The relevant hypotheses are $H_0 : p_s = p_r = p_g = p_p = 0.25$ vs H_a : Some $p_i \neq 0.25$, where the p_i's are the proportions in the Since we are only interested in cases where one performed higher on the math and verbal sections we ignore the students who scored the same on each. The relevant hypotheses are $H_0 : p_m = p_v = 0.5$ vs H_a : Some $p_i \neq 0.5$, where the p_m and p_v are the proportions with higher Math or Verbal SAT scores, respectively.

The total number of students (ignoring the ties) is $205 + 150 = 355$ so the expected count in each cell, assuming equally likely, is $355(0.5) = 177.5$. We compute a chi-square test statistic as

$$\chi^2 = (205 - 177.5)^2/177.5 + (150 - 177.5)^2/177.5 = 8.52$$

Using a the upper tail of chi-square distribution with 1 degree of freedom yields a p-value of 0.0035. At a 5% significance level, we conclude that students are not equally likely to have higher Math or Verbal scores. From the data, we see that students from this population are more likely to have a higher Math score. (Note that since there are only two categories after we eliminate the ties, we could have also done this problem as a z-test for a single proportion with $H_0 : p = 0.5$.)

D.31 This is a chi-square goodness-of-fit test.

(a) We see that the number of girls diagnosed with ADHD is $1960 + 2358 + 2859 + 2904 = 10,081$.

(b) The expected count for January to March is $n \cdot p_i = 10,081(0.243) = 2449.7$. We find the other expected counts similarly, shown in the table below.

Birth Date	Observed	Expected	Contribution to χ^2
Jan-Mar	1960	2449.7	97.9
Apr-Jun	2358	2600.9	22.7
Jul-Sep	2859	2590.8	27.8
Oct-Dec	2904	2439.6	88.4

(c) The contribution to the chi-square statistic for the January to March cell is

$$\frac{(observed - expected)^2}{expected} = \frac{(1960 - 2449.7)^2}{2449.7} = 97.9$$

This number and the contribution for each of the other cells, computed similarly, are shown in the table above, and the chi-square statistic is the sum: $\chi^2 = 97.9 + 22.7 + 27.8 + 88.4 = 236.8$.

(d) Since there are four categories, one for each quarter of the year, the degrees of freedom is $4 - 1 = 3$. The chi-square test statistic is very large (way out in the far reaches of the tail of the χ^2-distribution) so the p-value is essentially zero.

(e) There is *very* strong evidence that the distribution of ADHD diagnoses for girls differs from the proportions of births in each quarter. By comparing the observed and expected counts we see that younger children in a classroom (Oct-Dec) are diagnosed more frequently than we expect, while older children in a class (Jan-Mar) are diagnosed less frequently.

D.33 (a) To see if a chi-square distribution is appropriate, we find the expected count in each cell by multiplying the total for the row by the total for the column and dividing by the overall sample size ($n = 85$). These expected counts are summarized in the table below and we see that the smallest expected count of 5.2 (Rain, Fall) is (barely) greater than five, so a chi-square distribution is reasonable.

	Spring	Summer	Fall	Winter	Total
Rain	5.4	5.7	5.2	5.7	22
No Rain	15.6	16.3	14.8	16.3	63
Total	21	22	20	22	85

(b) The null hypothesis is that distribution of rain/no rain days in San Diego does not depend on the season and the alternative is that the rain/no rain distribution is related to the season. We have already calculated the expected counts in part (a) so we proceed to compute the chi-square test statistic by summing $(observed - expected)^2/expected$ for each cell

$$\chi^2 = \frac{(5 - 5.4)^2}{5.4} + \frac{(0 - 5.7)^2}{5.7} + \frac{(6 - 5.2)^2}{5.2} + \cdots + \frac{(11 - 16.3)^2}{16.3} = 14.6$$

Comparing $\chi^2 = 14.6$ to the upper tail of a chi-square distribution with 3 degrees of freedom yields a p-value of 0.002.

(c) This is a small p-value, so we reject the null hypothesis, indicating that there is a difference in the proportion of rainy days among the four seasons, and it appears the rainy season (with almost twice as many rainy days as expected if there were no difference) is the winter.

D.35 This is a chi-square test for association for a 4×2 table. The relevant hypotheses are
H_0 : Drinking habits are not related to gender
H_a : Drinking habits are related to gender

We compute the expected counts. For example, the expected count for males with 0 drinks is $(8956 \cdot 18712)/27268 = 6145.84$. The computer output below shows, for each cell, the observed counts with expected counts below them and contribution to the chi-square statistic below that.

```
              M          F   Total
  0         5402      13310   18712
          6145.84   12566.16
           90.027     44.030

  1 - 2     2147       3678    5825
          1913.18    3911.82
           28.575     13.976

  3 - 4      912        966    1878
           616.82    1261.18
          141.262     69.088

  5 +        495        358     853
           280.16     572.84
          164.744     80.573

Total       8956      18312   27268
```

Chi-Sq = 632.276, DF = 3, P-Value = 0.000

We see that the χ^2 test statistic is 632.276 and the p-value is essentially zero. This provides very strong evidence that drinking habits are not the same between males and females. We see that observed counts for males are less than expected for nondrinkers and greater than expected for more drinks, whereas that pattern is switched for females. Males tend to drink more alcoholic beverages than females.

D.37 (a) The null hypothesis is that all the state average home prices are equal and the alternative is that at least two states have different means.
$$H_0 : \mu_{CA} = \mu_{NY} = \mu_{NJ} = \mu_{PA}$$
$$H_a : \text{Some } \mu_1 \neq \mu_j$$

(b) We are comparing $k = 4$ states, so the numerator degrees of freedom is $k - 1 = 3$.

(c) The overall sample size is 120 homes, so the denominator degrees of freedom is $n - k = 116$.

(d) The sum of squares for error will tend to be much greater than the sum of squares for groups because we will be dividing the sum of squares for error by 116 to standardize it, while we will only be dividing the sum of squares groups by 3. However, without looking at the data, we cannot tell for sure and we have even less knowledge about how the *mean squares* might compare.

D.39 The null hypothesis is that the mean fiber amounts are all the same and the alternative hypothesis is that at least two of the companies have different mean amounts of fiber.
$$H_0 : \mu_{GM} = \mu_K = \mu_Q$$

$$H_a : \text{Some } \mu_1 \neq \mu_j$$

Since there are 3 groups (the three companies), the degrees of freedom for groups is 2. Since the sample size is 30, the total degrees of freedom is 29. The error degrees of freedom is $30 - 3 = 27$. We subtract to find the error sum of squares: $SSError = SSTotal - SSGroups = 102.47 - 4.96 = 97.51$. Filling in the rest of the ANOVA table, we have:

```
Source    DF      SS     MS     F      P
Company   2      4.96   2.48  0.69  0.512
Error     27    97.51   3.61
Total     29   102.47
```

The p-value of 0.512 is found using the upper tail (beyond $F = 0.69$) for an F-distribution with 2 numerator df and 27 denominator df. The p-value is very large so this sample does not provide evidence that the mean number of grams of fiber differs between the three companies.

D.41 (a) We use the summary data to calculate the various sums of squares

$$SSgroups = \sum n_i(\bar{x}_i - \bar{x})^2$$
$$= 31(249.05 - 143.59)^2 + 12(286.58 - 143.59)^2 + \cdots + 13(86.91 - 143.59)^2$$
$$= 956,816$$

$$SSE = \sum (n_i - 1)s_i^2$$
$$= (31 - 1)280.29^2 + (12 - 1)222.42^2 + \cdots + (13 - 1)58.95^2$$
$$= 3,768,795$$

$$SSTotal = (n - 1)s^2 = 130 \times 190.66^2 = 4,725,661$$

Up to round-off differences we see that

$$SStotal = 4,725,661 \approx SSGroups + SSE = 956,816 + 3,768,795$$

Note that we could also calculate any two of the quantities above and solve for the third.

We can now fill in the ANOVA table as follows to calculate the F-statistic:

Source	df	Sum of Sq.	Mean Square	F-statistic	p-value
Groups	$7 - 1 = 6$	956816	$\dfrac{956,816}{6} = 159,469$	$\dfrac{159,469}{30,393.5} = 5.25$	
Error	$131 - 7 = 124$	3,768,795	$\dfrac{3,768,795}{124} = 30,393.5$		
Total	$131 - 1 = 130$	4,725,661			

The F-statistic is 5.25.

(b) No, the ANOVA conditions are not satisfied. The sample sizes within each group are small, and the condition of normality is clearly violated as seen by the high outliers and skewness in the boxplots for several groups. Also, the condition of equal variability is not satisfied because the largest sample standard deviation, 280.29, is more than twice the smallest, 49.92. It would not be appropriate to use an F-distribution here.

D.43 In every case, we are testing $H_0 : \mu_i = \mu_j$ vs $H_a : \mu_i \neq \mu_j$ where μ_i and μ_j represent the mean temperature increase in the indicated conditions.

Testing legs together vs lap pad, we have

$$t = \frac{\overline{x}_i - \overline{x}_j}{\sqrt{MSE\left(\frac{1}{n_i} + \frac{1}{n_j}\right)}} = \frac{2.31 - 2.18}{\sqrt{0.63\left(\frac{1}{29} + \frac{1}{29}\right)}} = 0.62$$

We use a t-distribution with $df = n - k = 87 - 3 = 84$ to find the p-value. This is a two-tailed test so the p-value is $2(0.268) = 0.536$. This large p-value shows no convincing evidence that a lap pad affects the mean temperature increase when legs are together.

Testing legs together vs legs apart, we have

$$t = \frac{\overline{x}_i - \overline{x}_j}{\sqrt{MSE\left(\frac{1}{n_i} + \frac{1}{n_j}\right)}} = \frac{2.31 - 1.41}{\sqrt{0.63\left(\frac{1}{29} + \frac{1}{29}\right)}} = 4.32$$

We again use a t-distribution with $df = 84$ to find the p-value. The t-statistic is quite large in magnitude so, even after multiplying by 2 for the two-tailed test, the p-value is essentially zero. There is very strong evidence of a difference in mean temperature increase between keeping the legs together or legs apart.

Testing lap pad vs legs apart, we have

$$t = \frac{\overline{x}_i - \overline{x}_j}{\sqrt{MSE\left(\frac{1}{n_i} + \frac{1}{n_j}\right)}} = \frac{2.18 - 1.41}{\sqrt{0.63\left(\frac{1}{29} + \frac{1}{29}\right)}} = 3.69$$

We again use a t-distribution with $df = 84$ to find the p-value. This is a two-tailed test so the p-value is $2(0.0002) = 0.0004$. There is strong evidence of a difference in mean temperature increase between keeping the legs apart or using a lap pad with legs together. The mean temperature increase is lower when keeping the legs apart.

In summary, there is no significant difference in mean temperature increase between using a lap pad or not using when while keeping the legs together. However, the mean temperature increase is significantly less than in either of those conditions when keeping the legs apart.

D.45 The hypotheses are $H_0 : \rho = 0$ vs $H_a : \rho \neq 0$, where ρ is the correlation between happiness score and average sleep for all students. The t-statistic is:

$$t = \frac{r\sqrt{n-2}}{\sqrt{1-r^2}} = \frac{0.104\sqrt{251}}{\sqrt{1-(0.104^2)}} = 1.66$$

Using a t-distribution with $n - 2 = 253 - 2 = 251$ degrees of freedom, for this two-tailed test, we find a p-value of $2(0.0491) = 0.098$. At a 5% level, we do not find convincing evidence for a linear relationship between this measure of happiness and average number of hours slept at night for all students. The results are borderline, however, and are significant at a 10% level.

D.47 (a) The slope is $b_1 = 0.0620$. If the number of drinks goes up by 1, the predicted number of classes missed goes up by 0.0620.

(b) The t-statistic is 1.24 and the p-value is 0.215. We do not find convincing evidence that the number of alcoholic drinks helps to predict the number of missed classes.

(c) We see that $R^2 = 0.6\%$. The number of alcoholic drinks predicts very little (only 0.6%) of the variability in number of classes missed.

(d) The F-statistic is 1.55 and the p-value is 0.215. This model based on the number of drinks is not effective at predicting the number of classes missed.

D.49 There are several problems with the regression conditions, the most serious of which is the number of large outliers; points well above the line in the scatterplot with regression line. These also contribute to the right skew in the histogram of residuals, violating the normality condition. The residuals vs fits plot doesn't show roughly equal bands on either side of the zero mean, rather we again see the several large positive residuals that aren't balanced with similar sized negative residuals below the line. There is no clear curvature in the data, but the residual vs fits plot shows an interesting pattern as the most extreme negative residuals decrease in regular fashion – not a random scatter. We should be hesitant to use inference based on a linear model for these data (including the earlier exercise for these variables).

D.51 (a) The 95% confidence interval for the mean response is 1559.3 to 1667.9. We are 95% confident that the mean number of points for all players who make 400 free throws in a season is between 1559.3 and 1667.9.

(b) The 95% prediction interval for the response is 1241.2 to 1986.0. We are 95% confident that a player who makes 400 free throws in a season will have between 1241.2 and 1986.0 points for the season.

D.53 (a) The coefficient of *Age* is 0.08378. All else being equal, a person one year older will have a predicted percent body fat that is about 0.084 higher. The coefficient of *Abdomen* is 1.0327. If a person gains one centimeter on his or her abdomen circumference (with all other variables remaining the same), the predicted percent body fat goes up by 1.0327.

(b) The p-value from the ANOVA test is 0.000 so this model is effective at predicting percent body fat.

(c) We see that $R^2 = 75.7\%$ so 75.7% of the variability in body fat can be explained by the model using these nine explanatory variables.

(d) We see from the p-values for the individual slopes that *Abdomen* is the most significant variable in the model, with a p-value of 0.000, while *Neck* is the least significant with a p-value of 0.998.

(e) Two of the variables are significant at the 5% level: *Abdomen* and *Wrist*.

D.55 (a) This is an association between two quantitative variables, so can be tested with either a test for correlation or a test for slope in simple linear regression. Here we do a test for correlation. Let ρ be the true correlation between number of piercings and GPA, and our hypotheses are then

$$H_0 : \rho = 0$$
$$H_a : \rho \neq 0$$

Using technology, we find the sample correlation between *Piercings* and *GPA* to be $r = 0.079$. We can find the p-value using technology or using the formula. Using the formula, we see that the t-statistic is then

$$t = \frac{r \cdot \sqrt{n-2}}{\sqrt{1-r^2}} = \frac{0.079 \cdot \sqrt{343-2}}{\sqrt{1-(0.079^2)}} = 1.48$$

We compare this to a t-distribution with df $= n - 2 = 343 - 2 = 341$, and get a p-value of 0.143. The results are insignificant, and we do not have sufficient evidence that an association exists between the number of piercings and GPA of college students.

(b) Using technology, we fit the multiple regression model with GPA as the response variable and $Piercings$ and SAT as explanatory variables. The output is below.

```
Coefficients:
            Estimate Std. Error t value Pr(>|t|)
(Intercept) 1.6035944  0.2022610   7.928 3.22e-14 ***
Piercings   0.0207638  0.0091077   2.280   0.0232 *
SAT         0.0012604  0.0001653   7.625 2.47e-13 ***

Residual standard error: 0.3674 on 340 degrees of freedom
Multiple R-squared: 0.1514,    Adjusted R-squared: 0.1464
F-statistic: 30.33 on 2 and 340 DF,  p-value: 7.593e-13
```

The p-value for $Piercings$ of 0.0232 indicates that, after accounting for SAT score, the number of piercings is significantly associated with GPA.

Note: In part (b) we could also use $Piercings$ as the response with GPA and SAT as the predictors and the result (significance of the relationship between GPA and $Pirecings$ after accounting for SAT) would be identical.

D.57 We might start by considering the correlations of each of the potential explanatory variables with $Wins$. The three strongest correlations are with $Runs$ (r=0.785), RBI (r=0.758), and $Walks$ (r=0.533). We fit these three variables in a multiple regression model to predict $Wins$ to obtain the output below.

```
The regression equation is Wins = - 5.0 + 0.421 Runs - 0.314 RBI - 0.0021 Walks

Predictor       Coef  SE Coef      T     P
Constant       -5.04    12.53  -0.40 0.690
Runs          0.4215   0.1609   2.62 0.014
RBI          -0.3138   0.1622  -1.93 0.064
Walks       -0.00205  0.02692  -0.08 0.940

S = 6.73425   R-Sq = 66.4%   R-Sq(adj) = 62.6%

Analysis of Variance
Source          DF       SS      MS      F     P
Regression       3  2332.90  777.63  17.15 0.000
Residual Error  26  1179.10   45.35
```

The overall model is effective (ANOVA p-value≈ 0) and it explains 66.4% of the variability in $Wins$ for this sample. However, the t-test for the coefficient of $Walks$ is very large (0.940), indicating that $Walks$ is not very useful in this model (perhaps $Walks$ is strongly related to $Runs$ or RBI so that it offers no additional information about $Wins$ if they are already in the model).

This model might be improved by dropping $Walks$ and/or perhaps adding one of the other variables. When choosing a model from among many predictors there is rarely an absolute "best" choice that is optimal by all criteria. For this reason, there are several (even many) reasonable models that would be acceptable choices for this situation.

Final Essential Synthesis Solutions

E.1 *Dear Congressman Daniel Webster,*

You recently criticized the American Community Survey as being a random survey. However, the fact that it is a random survey is crucial *for enabling us to make generalizations from the sample of people surveyed to the entire population of US residents. We can only generalize from the sample to the population if the sample is representative of the population (closely resembles the population in all characteristics, except only smaller). Unfortunately, without randomness we are notoriously bad at choosing representative samples. Because the whole point of the survey is to gain information about the population, we do not know what the population looks like, and so have no way of knowing what is "representative". On the bright side,* randomly *choosing a sample yields a group of people that are representative of the population. With a random sample, the larger the sample size, the closer the sample statistics will be to the population values you care about. With non-random samples this may not be the case. In short, we can best draw valid scientific conclusions from samples that have been randomly selected.*

Sincerely,
A statistics student

E.3   (a) A bootstrap distribution, generated via StatKey and showing the cutoffs for the middle 90% is shown below:

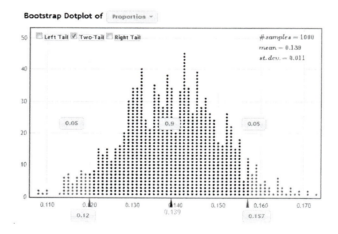

Based on the middle 90% of bootstrap proportions, a 90% confidence interval is 0.120 to 0.157.

(b) We estimate the standard error using the formula

$$SE = \sqrt{\frac{\hat{p}(1-\hat{p})}{n}} = \sqrt{\frac{0.139(1-0.139)}{1000}} = 0.011.$$

Notice that this matches the standard deviation of the bootstrap distribution. For a 90% confidence interval $z^* = 1.645$, so we generate the interval as

$$
\begin{array}{ccc}
\text{sample statistic} & \pm & z^* \cdot SE \\
\hat{p} & \pm & z^* \cdot SE \\
0.139 & \pm & 1.645 \cdot 0.011 \\
0.139 & \pm & 0.018 \\
0.121 & \text{to} & 0.157
\end{array}
$$

(c) We are 90% confident that between 12.1% and 15.7% of US residents do not have health insurance.

(d) The sample statistic is $\hat{p} = 0.139$ and the margin of error is $z^* \cdot SE = 1.645 \cdot 0.011 = 0.018$, or the distance from either bound of the confidence interval to the statistic, $0.157 - 0.139 = 0.018$.

(e) The sample size is much larger for the entire American Community Survey sample than for the 1000 people sub-sampled for **ACS**, so the margin of error will be much smaller.

(f) The 90% confidence interval is sample statistic $\pm$ margin of error, or $0.155 \pm 0.001 = (0.154, 0.156)$. Based on the full ACS survey, we are 90% confident that between 15.4% and 15.6% of all US residents do not have health insurance.

E.5 (a) *Income* is a quantitative variable, so we can visualize it's distribution with a histogram:

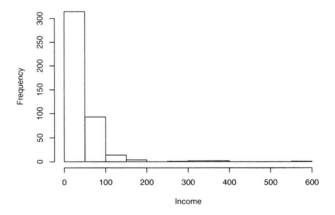

The distribution of income for employed US residents is strongly right-skewed. Most people have yearly incomes below \$100,000, but some people have incomes that are much higher. The maximum yearly income in this dataset is an outlier making about \$600,000 a year.

(b) The mean yearly income in the sample is $\overline{x} = \$41,494$, while the median income is \$29,000. The standard deviation of incomes is $s = \$52,248$ and the $IQR = \$40,900$. Yearly incomes in this dataset range from a minimum of \$0 to a maximum of \$563,000.

(c) We are looking at the relationship between a quantitative variable and a categorical variable, so can visualize with side-by-side boxplots:

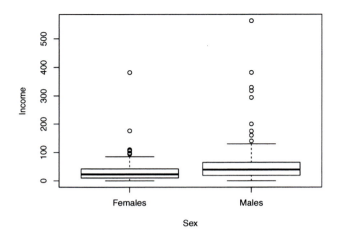

It appears that males tend to make more than females. The age distributions for each sex are heavily skewed towards larger incomes with several high outliers.

(d) In this sample, the males make an average of $\overline{x}_M = \$50,971$, while the females make an average of $\overline{x}_F = \$32,158$. The males make an average of $\$50,971 - \$32,158 = \$18,813$ more in income than the females.

(e) Let μ_M and μ_F denote the average yearly income for employed males and females, respectively, who live in the US. We test the hypotheses

$$H_0 : \mu_M = \mu_F$$
$$H_a : \mu_M \neq \mu_F.$$

We use StatKey or other technology to create a randomization distribution for difference in means, as shown below:

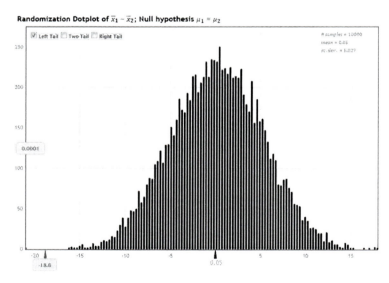

Of the 10,000 simulated randomization statistics, one one was less than the observed statistic (-18.8), so the area in one tail is 0.0001. We are testing a two-sided alternative, so double this to get a p-value of $2 \cdot 0.0001 = 0.0002$. This p-value is very small, so we have strong evidence that the average yearly income among employed US residents in 2010 is higher for males than for females.

E.7 (a) The number of people with health insurance in each racial group is the number of people of that race multiplied by the proportion of that race with health insurance. So the number of white people with health insurance is $761 \cdot 0.880 = 669.7$, which rounds to 670 (counts of people have to be whole numbers). The complete table is given below (and we could also generate this table directly from the **ACS** data).

	White	Black	Asian	Other
Health Insurance	670	86	59	46
No Health insurance	91	20	11	17

(b) This is a visualization of two categorical variables, which can be done with a segmented bar chart:

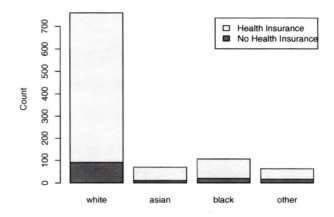

(c) We are testing for an association between two categorical variables, one of which has more than two levels, so use a chi-square test for association. The hypotheses are

H_0: There is no association between health insurance status and race
H_a: There is an association between health insurance status and race

We compute expected counts for each cell using (row total)(column total)/(sample size). The table of observed (expected) counts is given below:

	White	Black	Asian	Other	Total
Health Insurance	670 (655.2)	86 (91.27)	59 (60.3)	46 (54.2)	861
No Health insurance	91 (105.8)	20 (14.7)	11 (9.7)	17 (8.8)	139
Total	761	106	70	63	1000

This gives a chi-square statistic of

$$\chi^2 = \sum \frac{(observed - expected)^2}{expected} = \frac{(670 - 655.2)^2}{655.2} + \cdots + \frac{(17 - 8.8)^2}{8.8} = 13.79$$

The expected counts are all greater than 5, so we can compare this to a chi-square distribution with $(2-1)(4-1) = 2$ degrees of freedom, and find the p-value as the area above $\chi^2 = 13.79$. This gives a p-value of 0.0032. There is strong evidence for an association between whether or not a person has health insurance and race.

E.9   (a) *HoursWk* and *Income* are both quantitative variables, so we visualize with a scatterplot. Note: We could also switch the variables between the axes, unless we've read ahead to part (c).

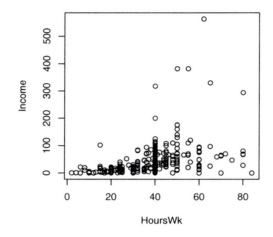

There appears to be a positive trend, although the association might be slightly curved, rather than linear, and the variability appears to increase as the number of hours worked increases.

   (b) These are both quantitative variables, so we do a test for correlation. The hypotheses are $H_0 : \rho = 0$ vs $H_a : \rho > 0$. The sample size is $n = 431$ and the sample correlation is $r = 0.379$. The relevant t-statistic is

$$t = \frac{r\sqrt{n-2}}{\sqrt{1-r^2}} = \frac{0.379\sqrt{431-2}}{\sqrt{1-0.379^2}} = 8.48.$$

We compare this to a t-distribution with $431 - 2 = 429$ degrees of freedom, and find the p-value is essentially 0. Hours worked per week and income are very significantly positively associated.

   (c) Some output for a regression model to predict *Income* based on *HoursWk* is given below:

```
Coefficients:
            Estimate Std. Error t value Pr(>|t|)
(Intercept) -18.3468     7.4349  -2.468    0.014 *
HoursWk       1.5529     0.1832   8.476 3.77e-16 ***

Residual standard error: 48.41 on 429 degrees of freedom
Multiple R-squared: 0.1435,    Adjusted R-squared: 0.1415
F-statistic: 71.85 on 1 and 429 DF,  p-value: 3.769e-16
```

The prediction equation is $\widehat{Income} = -18.3468 + 1.5529 \cdot HoursWk$.

   (d) The predicted yearly income for someone who works 40 hours a week is

$$\widehat{Income} = -18.3468 + 1.5529 \cdot 40 = 43.769$$

or about \$43,769.

(e) The percent of the variability in income explained by the number of hours worked per week is $R^2 = 14.35\%$.

(f) Below is a scatterplot with the regression line on it:

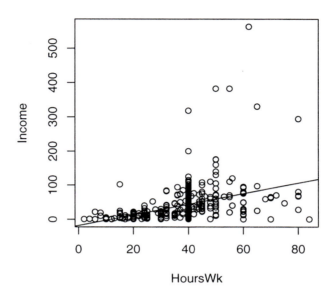

The condition of constant variability is clearly violated; variability in income is much higher when more hours are worked per week. There might also be a small amount of curvature in the relationship – note the relatively large number of points below the line for 20-35 hours of work a week and the negative predicted incomes when hours per week is very small.

E.11 Skull size is quantitative and which mound each skull was found in is categorical with two categories, so they should use a test for a difference in means.

E.13 We would like to make a statement about a population based on a sample proportion, so would use a confidence interval for a proportion.

E.15 We care about a single proportion (proportion of numbers ending in 0 or 5), and wish to determine whether this differs from 0.20, so would use a test for a proportion.

E.17 Both of these variables are quantitative, and the question is asking whether a negative correlation exists, so we would do a one-sided test for correlation. An lower-tail test for the slope in a regression model would also be acceptable.

E.19 This is testing for an association between two categorical variables, each with two categories, so we could use either a test for difference in proportions or a chi-square test for association.

E.21 We want to estimate a proportion, so would do an interval for a proportion.

E.23 We want to predict human equivalent age (quantitative) based on dog age (quantitative), so use slope of the simple linear regression.

E.25 Even though the experiment contains four groups, we are only interested in a categorical variable with two categories; exercise instructions or not, and one quantitative variable, amount of weight loss. The goal is not to test for a difference between the groups, but to estimate the difference between the groups, so we use a confidence interval for difference in means.

E.27 This involves two categorical variables (land or not, double loop or double flip), and she is interested whether the proportion of landing is higher for loop or flip, so should use a test for difference in proportions.

Section 11.1 Solutions

11.1 We have $P(\text{not } A) = 1 - P(A) = 1 - 0.4 = 0.6$.

11.3 By the additive rule, we have $P(A \text{ or } B) = P(A) + P(B) - P(A \text{ and } B) = 0.4 + 0.3 - 0.1 = 0.6$.

11.5 We have
$$P(B \text{ if } A) = \frac{P(A \text{ and } B)}{P(A)} = \frac{0.1}{0.4} = 0.25.$$

11.7 We need to check whether $P(A \text{ and } B) = P(A) \cdot P(B)$. Since $P(A \text{ and } B) = 0.1$ and $P(A) \cdot P(B) = 0.4 \cdot 0.3 = 0.12$, the events are not independent. We can also check from an earlier exercise that $P(B \text{ if } A) = 0.25 \neq P(B) = 0.3$.

11.9 We have $P(\text{not } B) = 1 - P(B) = 1 - 0.4 = 0.6$.

11.11 We have
$$P(A \text{ if } B) = \frac{P(A \text{ and } B)}{P(B)} = \frac{0.25}{0.4} = 0.625.$$

11.13 No! If A and B were disjoint, that means they cannot both happen at once, which means $P(A \text{ and } B) = 0$. Since we are told that $P(A \text{ and } B) = 0.25$, not zero, the events are not disjoint.

11.15 Since A and B are independent, knowing that B occurs gives us no additional information about A, so $P(A \text{ if } B) = P(A) = 0.7$.

11.17 Since A and B are independent, we have $P(A \text{ and } B) = P(A) \cdot P(B) = 0.7 \cdot 0.6 = 0.42$.

11.19 There are two cells that are included as part of event A, so $P(A) = 0.2 + 0.1 = 0.3$.

11.21 There is one cell that is in both event A and event B, so we have $P(A \text{ and } B) = 0.2$.

11.23 We have
$$P(A \text{ if } B) = \frac{P(A \text{ and } B)}{P(B)} = \frac{0.2}{0.6} = 0.333.$$

11.25 No! If A and B were disjoint, that means they cannot both happen at once, which means $P(A \text{ and } B) = 0$. We see in the table that $P(A \text{ and } B) = 0.2$, not zero, so the events are not disjoint.

11.27 The two events are disjoint, since if at least one skittle is red then all three can't be green. However, they are not independent or complements.

11.29 The two events are independent, as Australia winning their rugby match will not change the probability that Poland wins their chess match. However, they are not disjoint or complements.

11.31 (a) It will not necessarily be the case that EXACTLY 1 in 10 adults are left-handed for every sample. We can only conclude that approximately 10% will be left-handed in the "long run" (for very large samples).

(b) The three outcomes each have probability $\frac{1}{3}$ *only* if they are equally likely. This may not be the case for the results of baseball pitches.

(c) To find the probability of two consecutive 1's on independent dice rolls we should multiply the probabilities instead of adding them. Using the multiplicative rule, the probability that two consecutive rolls land with a 1 is $\frac{1}{6} \times \frac{1}{6} = \frac{1}{36}$.

(d) A probability that is not between 0 and 1 does not make sense.

11.33 (a) The description of the exercise indicates that $P(D) = 0.19$, $P(L) = 0.23$, and $P(D \text{ and } L) = 0.17$.

(b) We use the additive rule for finding the probability of one event *or* another.

$$P(D \text{ or } L) = P(D) + P(L) - P(D \text{ and } L) = 0.19 + 0.23 - 0.17 = 0.25$$

The probability that a country is either developed or large is 0.25.

(c) We find the probability that a country is developed *if* it is large, so we have

$$P(D \text{ if } L) = \frac{P(D \text{ and } L)}{P(L)} = \frac{0.17}{0.23} = 0.739$$

About 74% of large countries are developed, compared to 19% of all countries. Large countries are far more likely to be developed.

(d) We find the probability that a country is large *if* it is classified as developed, so we have

$$P(L \text{ if } D) = \frac{P(D \text{ and } L)}{P(D)} = \frac{0.17}{0.19} = 0.895$$

About 90% of developed countries are large, compared to 23% of all countries. Developed countries are far more likely to be large.

(e) To find the probability a country is not developed we have $P(\text{not } D) = 1 - P(D) = 1 - 0.19 = 0.81$.

(f) Saying D and L are disjoint would mean that there are no countries that are both developed and large. This is not true since we are told that 17% of all countries are both developed and large. It is clear that D and L are not disjoint.

(g) Saying D and L are independent would mean that knowing a country is large would give us no information about whether it is developed or not and knowing it is developed would give us no information about whether it is large or not. This is clearly not true, as parts (c) and (d) show, the conditional probabilities are much different than the probabilities without the condition. These two events are not at all independent.

11.35 (a) We are finding $P(C)$. There are a total of 251 inductees and 227 of them were born in Canada, so we have

$$P(C) = \frac{227}{251} = 0.90$$

The probability is 0.90 that an inductee selected at random is Canadian. It is remarkable how much Canada has dominated the sport!

(b) We are finding $P(\text{not } D)$. There are a total of 251 inductees and 77 of them play defense while the other 174 do not, so we have

$$P(\text{not } D) = \frac{174}{251} = 0.69$$

The probability that an inductee selected at random will not be a defenseman is 0.69.

(c) We are finding $P(C$ and $D)$. Of the 251 inductees, there are 71 that are defensemen born in Canada, so we have

$$P(C \text{ and } D) = \frac{71}{251} = 0.28$$

(d) We are finding $P(C$ or $D)$. Of the 251 inductees, 227 were born in Canada and 77 are defensemen and 71 are both, so we have

$$P(C \text{ or } D) = P(C) + P(D) - P(C \text{ and } D) = \frac{227}{251} + \frac{77}{251} - \frac{71}{251} = \frac{233}{251} = 0.93$$

(e) In this case, we are interested only in inductees who are Canadian, and we want to know the probability that a Canadian inductee plays defense, $P(D$ if $C)$. There are 227 Canadians and 71 of them play defense, so we have

$$P(D \text{ if } C) = \frac{71}{227} = 0.31$$

(f) In this case, we are interested only in defensemen, and we want to know the probability that a defenseman inductee is also Canadian, $P(C$ if $D)$. There are 77 defensemen and 71 of them are Canadian, so we have

$$P(C \text{ if } D) = \frac{71}{77} = 0.92$$

11.37 (a) There are 18 yellow ones out of a total of 80, so the probability that we pick a yellow one is $18/80 = 0.225$.

(b) The probability that it *is* brown is $8/80 = 0.10$, so the probability that it is not brown is $1-0.10 = 0.90$.

(c) The single piece can be blue or green, but not both, so these are disjoint events. The probability the randomly selected candy is blue or green is $20/80 + 11/80 = 31/80 = 0.3875$.

(d) The probability that the first one is red is $11/80 = 0.1375$. When we put it back and mix them up, the probability that the next one is red is also 0.1375. By the multiplication rule, since the two selections are independent, the probability both selections are red is $0.1375 \cdot 0.1375 = 0.0189$.

(e) The probability that the first one is yellow is $18/80$. Once that one is taken (since we don't put it back and we eat it instead), there are only 79 pieces left and 20 of those are blue. By the multiplication rule, the probability is $(18/80) \cdot (20/79) = 0.0570$.

11.39 Let CBM and CBW denote the events that a man or a woman is colorblind, respectively.

(a) As 7% of men are colorblind, $P(CBM) = 0.07$.

(b) As 0.4% of women are colorblind, $P(\text{not } CBW) = 1 - P(CBW) = 1 - 0.004 = 0.996$.

(c) The probability the woman is not colorblind is 0.996, and the probability that the man is not colorblind is $1 - 0.07 = 0.93$. As the man and woman are selected independently, we can multiply their probabilities:

$$P(\text{Neither is Colorblind}) = P(\text{not } CBM) \cdot P(\text{not } CBW) = 0.93 \times 0.996 = 0.926.$$

(d) The event that "At least one is colorblind" is the complement of part (d) that "Neither is Colorblind" so we have

$$P(\text{At least one is Colorblind}) = 1 - P(\text{Neither is Colorblind}) = 1 - 0.926 = 0.074$$

We could also do this part as

$$P(CBM \text{ or } CBW) = P(CBM) + P(CBW) - P(CBM \text{ and } CBW) = 0.07 + 0.004 - (0.07)(0.004) = 0.074$$

11.41   (a) As 85,227 out of 100,000 men live to age 60, the probability is $\frac{85227}{100000} = 0.8523$.

  (b) As 72,066 out of 100,000 men live to age 70, the probability is $1 - \frac{72066}{100000} = 0.2793$.

  (c) As 15,722 out of 100,000 men live to age 90 and 12,986 out of 100,000 live to age 91, the probability that a man dies at age 90 is $\frac{15722 - 12986}{100000} = 0.0274$.

  (d) Use conditional probability:

$$P(\text{dies at 90 if lives till 90}) = \frac{P(\text{dies at 90 and lives till 90})}{P(\text{lives till 90})} = \frac{P(\text{dies at 90})}{P(\text{lives till 90})} = \frac{0.0274}{15722/100000} = 0.1743.$$

  (e) Use conditional probability:

$$P(\text{dies at 90 if lives till 80}) = \frac{P(\text{dies at 90 and lives till 80})}{P(\text{lives till 80})} = \frac{P(\text{dies at 90})}{P(\text{lives till 80})} = \frac{0.0274}{47974/100000} = 0.0571.$$

  (f) As 85,227 out of 100000 men live to age 60 and 15,722 out of 100000 live to age 90, the probability that a man dies between the ages of 60 and 89 is $\frac{85227 - 15722}{100000} = 0.6951$.

  (g) Use conditional probability:

$$P(\text{lives till 90 if lives till 60}) = \frac{P(\text{lives till 90 and lives till 60})}{P(\text{lives till 60})} = \frac{P(\text{lives till 90})}{P(\text{lives till 60})} = \frac{15722/100000}{85227/100000} = 0.1845.$$

11.43 If you are served one of the pancakes at random, let A be the event that the side facing you is burned and B be the event that the other side is burned. We want to find $P(B \text{ if } A)$. As only one of three pancakes is burned on both sides, $P(A \text{ and } B) = 1/3$. As 3 out of 6 total sides are burned, $P(A) = 3/6 = 1/2$. So,

$$P(B \text{ if } A) = \frac{P(A \text{ and } B)}{P(A)} = \frac{1/3}{1/2} = \frac{2}{3}$$

Section 11.2 Solutions

11.45 Here is the tree with the missing probabilities filled in.

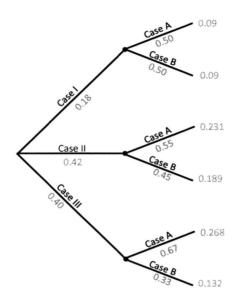

We use the conditional probability rule to find

$$P(A \text{ if } I) = \frac{P(I \text{ and } A)}{P(I)} = \frac{0.09}{0.18} = 0.5$$

We use the complement rule (sum of a branches from one point equals one) to find

$$P(B \text{ if } I) = 1 - P(A \text{ if } I) = 1 - 0.5 = 0.5$$

We use the multiplicative rule to find

$$P(I \text{ and } B) = P(I) \cdot P(B \text{ if } I) = 0.18(0.5) = 0.09$$

From the multiplicative rule for III and A we have

$$P(III \text{ and } A) = P(III) \cdot P(A \text{ if } III) \implies 0.268 = P(III) \cdot 0.67 \implies P(III) = \frac{0.268}{0.67} = 0.4$$

Using the complement rule several more times

$$\begin{aligned}
P(II) &= 1 - P(I) - P(III)) = 1 - 0.18 - 0.4 = 0.42 \\
P(A \text{ if } II) &= 1 - P(B \text{ if } II) = 1 - 0.45 = 0.55 \\
P(B \text{ if } III) &= 1 - P(A \text{ if } III) = 1 - 0.67 = 0.33
\end{aligned}$$

Finally, several more multiplicative rules along pairs of branches give

$$
\begin{aligned}
P(II \text{ and } A) &= P(II) \cdot P(A \text{ if } II) = 0.42(0.55) = 0.231 \\
P(II \text{ and } B) &= P(II) \cdot P(B \text{ if } II) = 0.42(0.45) = 0.189 \\
P(III \text{ and } B) &= P(III) \cdot P(B \text{ if } III) = 0.40(0.33) = 0.132
\end{aligned}
$$

11.47 Here is the tree with the missing probabilities filled in.

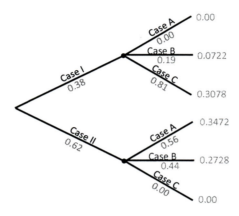

Using the complement rule (sum of all branches from one point must be one), we have

$$
\begin{aligned}
P(II) &= 1 - P(I) = 1 - 0.38 = 0.62 \\
P(C \text{ if } I) &= 1 - P(A \text{ if } I) - P(B \text{ if } I) = 1 - 0.00 - 0.19 = 0.81
\end{aligned}
$$

By the conditional probability rule , we have

$$
P(C \text{ if } II) = \frac{P(II \text{ and } C)}{P(II)} = \frac{0.00}{0.62} = 0.00
$$

One more complement rule gives

$$
P(B \text{ if } II) = 1 - P(A \text{ if } II) - P(C \text{ if } II) = 1 - 0.56 - 0.00 = 0.44
$$

Finally, we apply the multiplicative rule several times to get

$$
\begin{aligned}
P(I \text{ and } A) &= P(I) \cdot P(A \text{ if } I) = 0.38(0.00) = 0.00 \\
P(I \text{ and } B) &= P(I) \cdot P(B \text{ if } I) = 0.38(0.19) = 0.0722 \\
P(I \text{ and } C) &= P(I) \cdot P(C \text{ if } I) = 0.38(0.81) = 0.3078 \\
P(II \text{ and } A) &= P(II) \cdot P(A \text{ if } II) = 0.62(0.56) = 0.3472 \\
P(II \text{ and } B) &= P(II) \cdot P(B \text{ if } II) = 0.62(0.44) = 0.2728
\end{aligned}
$$

11.49 We use the multiplicative rule to see $P(A \text{ and } S) = P(A) \cdot P(S \text{ if } A) = 0.6 \cdot 0.1 = 0.06$.

11.51 This conditional probability is shown directly on the tree diagram. Since we assume B is true, we follow the B branch and then find the probability of S, which we see is 0.8.

11.53 We see in the tree diagram that there are two ways for S to occur. We find these two probabilities and add them up, using the total probability rule. We see that $P(A \text{ and } S) = 0.6 \cdot 0.1 = 0.06$ so this branch can be labeled 0.06. We also see that $P(B \text{ and } S) = 0.4 \cdot 0.8 = 0.32$. Since either A or B must occur (since these are the only two branches in this part of the tree), these are the only two ways that S can occur, so $P(S) = P(A \text{ and } S) + P(B \text{ and } S) = 0.06 + 0.32 = 0.38$.

11.55 We know that

$$P(B \text{ if } R) = \frac{P(B \text{ and } R)}{P(R)}$$

so we need to find $P(B \text{ and } R)$ and $P(R)$. Using the multiplicative rule, we see that $P(B \text{ and } R) = 0.4 \cdot 0.2 = 0.08$. We also see that $P(A \text{ and } R) = 0.6 \cdot 0.9 = 0.54$. By the total probability rule, we have

$$P(B \text{ if } R) = \frac{P(B \text{ and } R)}{P(R)} = \frac{P(B \text{ and } R)}{P(A \text{ and } R) + P(B \text{ and } R)} = \frac{0.08}{0.54 + 0.08} = 0.129$$

We see that $P(B \text{ if } R) = 0.129$.

11.57 We use F to denote the event of having fibromyalgia and R to denote the event of having restless leg syndrome. The tree diagram is shown, and we use the multiplication rule to find the probabilities at the end of the branches. For example, for the top branch, we have $P(F \text{ and } R) = 0.02 \cdot 0.33 = 0.0066$.

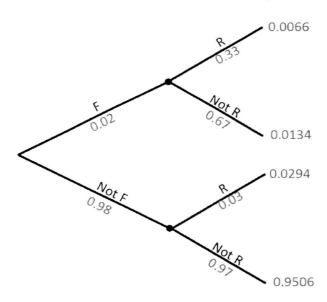

We are finding $P(F \text{ if } R)$. By adding the probabilities of the R branches, we see that $P(R) = 0.0066 + 0.0294 = 0.036$. The conditional probability is

$$P(F \text{ if } R) = \frac{P(F \text{ and } R)}{P(R)} = \frac{0.0066}{0.036} = 0.183.$$

If a person has restless leg syndrome, the probability that the person will also have fibromyalgia is about 18%.

11.59 Let F, C, and S denote the three types of pitches (fastball, curveball, spitball) and K denote the pitch is a strike. We are given the probability for each type of pitch

$$P(F) = 0.60 \qquad P(C) = 0.25 \qquad P(S) = 1 - P(F) - P(C) = 1 - 0.60 - 0.25 = 0.15$$

and the conditional probability for a strike with each pitch
$$P(K \text{ if } F) = 0.70 \qquad P(K \text{ if } C) = 0.50 \qquad P(K \text{ if } S) = 0.30$$
We need to find the probability of a certain type of pitch (curveball) given that we know the outcome (strike). This calls for an application of Bayes' rule:

$$
\begin{aligned}
P(C \text{ if } K) &= \frac{P(C)P(K \text{ if } C)}{P(F)P(K \text{ if } F) + P(C)P(K \text{ if } C) + P(S)P(K \text{ if } S)} \\
&= \frac{0.25 \cdot 0.50}{0.60 \cdot 0.70 + 0.25 \cdot 0.50 + 0.15 \cdot 0.30} \\
&= \frac{0.125}{0.59} \\
&= 0.212
\end{aligned}
$$

If we see Slippery throw a strike, there's a 21.2% chance it was with a curveball.

11.61 Using the Bayes' rule,

$$P(\text{Spam if Text}) = \frac{P(\text{Spam})P(\text{Text if Spam})}{P(\text{Text})} = \frac{0.134 \cdot 0.3855}{0.0701} = 0.737$$

11.63 Applying the total probability rule,

$$
\begin{aligned}
P(\text{Free and Text}) &= P(\text{Spam})P(\text{Text and Free if Spam}) + P(\text{not Spam})P(\text{Text and Free if not Spam}) \\
&= 0.134 \cdot 0.17 + 0.866 \cdot 0.0006 \\
&= 0.0233.
\end{aligned}
$$

Again applying the total probability rule,

$$P(\text{Free and not Text}) = P(\text{Free}) - P(\text{Free and Text}) = 0.0475 - 0.0233 = 0.0242$$

and

$$P(\text{Free and not Text if Spam}) = P(\text{Free if Spam}) - P(\text{Free and Text if Spam}) = 0.2664 - 0.1700 = 0.0964.$$

So, applying Bayes' rule,

$$P(\text{Spam if Free and not Text}) = \frac{P(\text{Spam})P(\text{Free and not Text if Spam})}{P(\text{Free and not Text})} = \frac{0.134 \cdot 0.0964}{0.0242} = 0.534.$$

Section 11.3 Solutions

11.65 A discrete random variable, as it can only take the values $\{0, 0.1, 0.2, \ldots, 1\}$).

11.67 Not a random variable, as it does not have a numerical outcome.

11.69 There are two conditions for a probability function. The first is that all the probabilities are between 0 and 1 and that is true here. The second is that the probabilities add up to 1.0, and that is true here because $0.4 + 0.3 + 0.2 + 0.1 = 1.0$.

11.71 We have $P(X > 1) = P(X = 2 \text{ or } X = 3 \text{ or } X = 4) = 0.3 + 0.2 + 0.1 = 0.6$.

11.73 We have $P(X \text{ is an odd number}) = P(X = 1 \text{ or } X = 3) = 0.4 + 0.2 = 0.6$.

11.75 The probability values have to add up to 1.0 so we have $? = 1.0 - (0.1 + 0.1 + 0.2) = 0.6$.

11.77 The probability values have to add up to 1.0 but the sum of the two values there is already greater than 1 (we see $0.5 + 0.6 = 1.1$). Since a probability cannot be a negative number, this cannot be a probability function.

11.79   (a) We multiply the values of the random variable (in this case, 1, 2, and 3) by the corresponding probability and add up the results. We have

$$\mu = 1(0.2) + 2(0.3) + 3(0.5) = 2.3$$

The mean of this random variable is 2.3.

  (b) To find the standard deviation, we subtract the mean of 2.3 from each value, square the difference, multiply by the probability, and add up the results to find the variance; then take a square root to find the standard deviation.

$$\begin{aligned} \sigma^2 &= (1 - 2.3)^2 \cdot 0.2 + (2 - 2.3)^2 \cdot 0.3 + (3 - 2.3)^2 \cdot 0.5 = 0.61 \\ \implies \sigma &= \sqrt{0.61} = 0.781 \end{aligned}$$

11.81   (a) We multiply the values of the random variable (in this case, 20, 30, 40, and 50) by the corresponding probability and add up the results. We have

$$\mu = 20(0.6) + 30(0.2) + 40(0.1) + 50(0.1) = 27$$

The mean of this random variable is 27.

  (b) To find the standard deviation, we subtract the mean of 27 from each value, square the difference, multiply by the probability, and add up the results to find the variance; then take a square root to find the standard deviation.

$$\begin{aligned} \sigma^2 &= (20 - 27)^2 \cdot 0.6 + (30 - 27)^2 \cdot 0.2 + (40 - 27)^2 \cdot 0.1 + (50 - 27)^2 \cdot 0.1 = 101 \\ \implies \sigma &= \sigma = \sqrt{101} = 10.05 \end{aligned}$$

11.83   (a) We see that $0.217 + 0.363 + 0.165 + 0.145 + 0.067 + 0.026 + 0.018 = 1.001$. This is different from 1 just by round-off error on the individual probabilities.

  (b) We have $p(1) + p(2) = 0.217 + 0.363 = 0.580$.

(c) We have $p(5) + p(6) + p(7) = 0.067 + 0.026 + 0.018 = 0.111$.

(d) It is easiest to find this probability using the complement rule, since more than 1 occupant is the complement of 1 occupant for this random variable. The answer is $1 - p(1) = 1 - 0.217 = 0.783$.

11.85   (a) We multiply the values of the random variable by the corresponding probability and add up the results. We have

$$\mu = 1(0.217) + 2(0.363) + 3(0.165) + 4(0.145) + 5(0.067) + 6(0.026) + 7(0.018) = 2.635$$

The average household size for an owner-occupied housing unit in the US is 2.635 people.

(b) To find the standard deviation, we subtract the mean of 2.635 from each value, square the difference, multiply by the probability, and add up the results to find the variance; then take a square root to find the standard deviation.

$$\begin{aligned}
\sigma^2 &= (1 - 2.635)^2 \cdot 0.217 + (2 - 2.635)^2 \cdot 0.363 + \cdots + (7 - 2.635)^2 \cdot 0.018 \\
&= 2.03072 \\
\Longrightarrow \sigma &= \sqrt{2.03072} = 1.425
\end{aligned}$$

11.87 Let the random variable X measure fruit fly lifetimes (in months).

(a) The probabilities must add to 1, so the proportion of dying in the second month is

$$P(X = 2) = 1 - (0.30 + 0.20 + 0.15 + 0.10 + 0.05) = 1 - 0.80 = 0.20$$

(b) $P(X > 4) = P(X = 5) + P(X = 6) = 0.10 + 0.05 = 0.15$

(c) The mean fruit fly lifetime is

$$\mu = 1(0.30) + 2(0.20) + 3(0.20) + 4(0.15) + 5(0.10) + 6(0.05) = 2.7 \text{ months}$$

(d) The standard deviation of fruit fly lifetimes is

$$\sigma = \sqrt{(1 - 2.7)^2 \cdot 0.30 + (2 - 2.7)^2 \cdot 0.20 + \cdots + (6 - 2.7)^2 \cdot 0.05} = \sqrt{2.31} = 1.52 \text{ months}$$

11.89   (a) This is a conditional probability, $P(A \text{ if } B)$, where the event A is $X = 1$ and the conditioning event B is $\{X = 1 \text{ or } X = 2\}$. From the probability function we see that $P(A) = 0.30$ and $P(B) = 0.30 + 0.20 = 0.50$. Note also that $P(A \text{ and } B) = P(A)$ since $X = 1$ is the only outcome they have in common. We have

$$P(A \text{ if } B) = \frac{P(A \text{ and } B)}{P(B)} = \frac{P(A)}{P(B)} = \frac{0.30}{0.50} = 0.6$$

(b) This is a conditional probability, $P(C \text{ if } D)$, where the event C is $\{X = 5 \text{ or } X = 6\}$ and the conditioning event D is $X \geq 3$. From the probability function we see that $P(C) = 0.10 + 0.05 = 0.15$ and $P(D) = 0.20 + 0.15 + 0.10 + 0.05 = 0.50$. Note also that $P(C \text{ and } D) = P(C)$ since $X = 5$ and $X = 6$ are the only outcomes they have in common. We have

$$P(C \text{ if } D) = \frac{P(C \text{ and } D)}{P(D)} = \frac{P(C)}{P(D)} = \frac{0.15}{0.50} = 0.3$$

A fruit fly that makes it safely through it's first two months will have a 30% chance of living to five or six months.

11.91   (a) There are three possible values for the random variable, $\{0,1,2\}$. Let S denote he successfully makes a shot and F that he fails to make it.

$$
\begin{aligned}
P(X = 2) &= P(S_1 \text{ and } S_2) = P(S_1) \cdot P(S_2) = 0.881 \cdot 0.881 = 0.776 \\
P(X = 0) &= P(F_1 \text{ and } F_2) = P(F_1) \cdot P(F_2) = 0.119 \cdot 0.119 = 0.014 \\
P(X = 1) &= P(F_1 \text{ and } S_2) + P(S_1 \text{ and } F_2) = 0.119 \cdot 0.881 + 0.881 \cdot 0.119 = 0.210
\end{aligned}
$$

So, the probability distribution of X is

x	0	1	2
$p(x)$	0.014	0.210	0.776

(b) The mean number of free throws made in two attempts is

$$\mu = 0(0.014) + 1(0.210) + 2(0.776) = 1.762$$

11.93   (a) We know that $P(X = \$29.95) = 2 \cdot P(X = \$39.95)$ and $P(X = \$23.95) = 3 \cdot P(X = \$39.95)$. It follows that

$$1 = P(X = \$23.95) + P(X = \$29.95) + P(X = \$39.95) = 6 \cdot P(X = \$39.95),$$

so $P(X = \$39.95) = \frac{1}{6}$. The probability distribution of X is

x	\$23.95	\$29.95	\$39.95
$p(x)$	3/6	2/6	1/6

(b) The mean μ of X is

$$\mu = \$23.95 \cdot \frac{3}{6} + \$29.95 \cdot \frac{2}{6} + \$39.95 \cdot \frac{1}{6} = \$28.62$$

(c) The variance σ^2 of X is

$$\sigma^2 = (23.95 - 28.62)^2 \cdot \frac{3}{6} + (29.95 - 28.62)^2 \cdot \frac{2}{6} + (39.95 - 28.62)^2 \cdot \frac{1}{6} = 32.89.$$

So, the standard deviation is $\sigma = \sqrt{32.89} = \$5.73$

11.95   (a) Using the formula for the probability function with $p = 1/6$ and $k = 3$ we have

$$P(X = 3) = \left(\frac{1}{6}\right)\left(1 - \frac{1}{6}\right)^{3-1} = \left(\frac{1}{6}\right)\left(\frac{5}{6}\right)^2 = 0.116$$

(b) The event "more than three turns to finish" or $X > 3$ includes $X = 4, 5, 6, \ldots$, an infinite number of possible outcomes! Fortunately we can use the complement rule.

$$
\begin{aligned}
P(X > 3) &= 1 - (p(1) + p(2) + p(3)) \\
&= 1 - \left[\left(\frac{1}{6}\right)\left(\frac{5}{6}\right)^0 + \left(\frac{1}{6}\right)\left(\frac{5}{6}\right)^1 + \left(\frac{1}{6}\right)\left(\frac{5}{6}\right)^2\right] \\
&= 1 - [0.1667 + 0.1389 + 0.1157] \\
&= 1 - 0.4213 = 0.5787
\end{aligned}
$$

Section 11.4 Solutions

11.97 Not binomial, since the number of rolls is not fixed.

11.99 This is a binomial random variable, with $n = 75$ and $p = 0.3$.

11.101 We see that $4! = 4 \cdot 3 \cdot 2 \cdot 1 = 24$.

11.103 We see that $8! = 8 \cdot 7 \cdot 6 \cdot 5 \cdot 4 \cdot 3 \cdot 2 \cdot 1 = 40,320$.

11.105 We have $\binom{8}{3} = \dfrac{8!}{3!(5!)} = \dfrac{8 \cdot 7 \cdot 6 \cdot 5 \cdot 4 \cdot 3 \cdot 2 \cdot 1}{3 \cdot 2 \cdot 1 \cdot 5 \cdot 4 \cdot 3 \cdot 2 \cdot 1} = \dfrac{336}{6} = 56$.

11.107 We have $\binom{10}{8} = \dfrac{10!}{8!(2!)} = \dfrac{10 \cdot 9 \cdot 8 \cdot 7 \cdot 6 \cdot 5 \cdot 4 \cdot 3 \cdot 2 \cdot 1}{8 \cdot 7 \cdot 6 \cdot 5 \cdot 4 \cdot 3 \cdot 2 \cdot 1 \cdot 2 \cdot 1} = \dfrac{90}{2} = 45$.

11.109 We first calculate that $\binom{6}{2} = \dfrac{6!}{2!(4!)} = 15$. We then find

$$P(X = 2) = \binom{6}{2}(0.3^2)(0.7^4) = 15(0.3^2)(0.7^4) = 0.324.$$

11.111 We first calculate that $\binom{10}{3} = \dfrac{10!}{3!(7!)} = 120$. We then find

$$P(X = 3) = \binom{10}{3}(0.4^3)(0.6^7) = 120(0.4^3)(0.6^7) = 0.215.$$

11.113 The mean is $\mu = np = 6(0.4) = 2.4$ and the standard deviation is

$$\sigma = \sqrt{np(1 - p)} = \sqrt{6(0.4)(0.6)} = \sqrt{1.44} = 1.2$$

11.115 The mean is $\mu = np = 30(0.5) = 15$ and the standard deviation is

$$\sigma = \sqrt{np(1 - p)} = \sqrt{30(0.5)(0.5)} = \sqrt{7.5} = 2.74$$

11.117 A probability function gives the probability for each possible value of the random variable. This is a binomial random variable with $n = 3$ and $p = 0.49$ (since we are counting the number of girls not boys). The probability of 0 girls is:

$$P(X = 0) = \binom{3}{0}(0.49^0)(0.51^3) = 1 \cdot 1 \cdot 0.51^3 = 0.133$$

The probability of 1 girl is:

$$P(X = 1) = \binom{3}{1}(0.49^1)(0.51^2) = 3 \cdot (0.49^1)(0.51^2) = 0.382$$

The probability of 2 girls is:

$$P(X = 2) = \binom{3}{2}(0.49^2)(0.51^1) = 3 \cdot (0.49^2)(0.51^1) = 0.367$$

The probability of 3 girls is:

$$P(X = 3) = \binom{3}{3}(0.49^3)(0.51^0) = 1 \cdot (0.49^3) \cdot 1 = 0.118$$

We can summarize these results with a table for the probability function.

x	0	1	2	3
$p(x)$	0.133	0.382	0.367	0.118

Notice that the four probabilities add up to 1, as we expect for a probability function.

11.119 If X is the random variable giving the number of college graduates in a random sample of 12 US adults, then X is a binomial random variable with $n = 12$ and $p = 0.275$. We are finding $P(X = 6)$. We first calculate

$$\binom{12}{6} = \frac{12!}{6!(6!)} = 924$$

We then find

$$P(X = 6) = \binom{12}{6}(0.275^6)(0.725^6) = 924(0.275^6)(0.725^6) = 0.058.$$

The probability is only 0.058 that exactly half of the sample are college graduates.

11.121 If X is the random variable giving the number of owner-occupied units in a random sample of 20 housing units in the US, then X is a binomial random variable with $n = 20$ and $p = 0.65$.

(a) To find $P(X = 15)$, we first calculate $\binom{20}{15} = \frac{20!}{15!(5!)} = 15,504$. We then find

$$P(X = 15) = \binom{20}{15}(0.65^{15})(0.35^5) = 15,504(0.65^{15})(0.35^5) = 0.1272.$$

(b) We know that $P(X \geq 18) = P(X = 18) + P(X = 19) + P(X = 20)$, and we calculate each of the terms separately and add them up. We have

$$P(X = 18) = \binom{20}{18}(0.65^{18})(0.35^2) = 190(0.65^{18})(0.35^2) = 0.0100$$

$$P(X = 19) = \binom{20}{19}(0.65^{19})(0.35^1) = 20(0.65^{19})(0.35^1) = 0.0020$$

$$P(X = 20) = \binom{20}{20}(0.65^{20})(0.35^0) = 1 \cdot (0.65^{20}) \cdot 1 = 0.0002$$

Then we have

$$P(X \geq 18) = P(X = 18) + P(X = 19) + P(X = 20) = 0.0100 + 0.0020 + 0.0002 = 0.0122$$

11.123 The mean is $\mu = np = 4(0.25) = 1$ senior (1 senior among the 4 students makes sense since about 25% of the overall student body are seniors).

The standard deviation is $\sigma = \sqrt{np(1-p)} = \sqrt{4(0.25)(0.75)} = \sqrt{0.75} = 0.866$.

11.125 This is a binomial random variable with $n = 10$ and $p = 0.13$.

The mean is $\mu = np = 10(0.13) = 1.3$ senior citizens in a sample of 10.

The standard deviation is $\sigma = \sqrt{np(1-p)} = \sqrt{10(0.13)(0.87)} = \sqrt{1.131} = 1.063$.

11.127   (a) Let X be the number of free throws Ray Allen makes during the game. Then, X has a binomial distribution with $n = 8$ and $p = 0.881$. So,

$$P(X \geq 7) = P(X = 7) + P(X = 8)$$
$$= \binom{8}{7} 0.881^7 0.119 + \binom{8}{8} 0.881^8 0.119^0$$
$$= 0.3922 + 0.3629$$
$$= 0.7551.$$

(b) Let X be the number of free throws Ray Allen makes during the season. Then, X has a binomial distribution with $n = 80$ and $p = 0.881$. So,

$$P(X \geq 70) = P(X = 70) + P(X = 71) + \ldots + P(X = 80)$$
$$= \binom{80}{70} 0.881^{70} 0.119^{10} + \binom{80}{71} 0.881^{71} 0.119^9 + \ldots + \binom{80}{80} 0.881^{80} 0.119^0$$
$$= 0.1319 + 0.1376 + \ldots + 0.00004$$
$$= 0.6469.$$

(c) The mean for $n = 8$ free throws in a game

$$\mu = 8 \cdot 0.881 = 7.05$$

The standard deviation for 8 free throws in a game is

$$\sigma = \sqrt{8 \cdot 0.881 \cdot 0.119} = 0.916$$

(d) The mean is for 80 free throws in the playoffs is

$$\mu = 80 \cdot 0.881 = 70.5$$

The standard deviation for 80 free throws in the playoffs is

$$\sigma = \sqrt{80 \cdot 0.881 \cdot 0.119} = 2.90$$

11.129 Let X be a binomial random variable with n trials and probability of success p and $Y = \frac{X}{n}$ be the corresponding proportion. So, $P(X = k) = P(Y = k/n)$ for all k. Let u be the mean of Y. As the mean of X is $np = \sum k \cdot P(X = k)$,

$$np = \sum k \cdot P(X = k)$$
$$np = n \left(\sum \frac{k}{n} \cdot P(X = k) \right)$$
$$np = n \left(\sum \frac{k}{n} \cdot P(Y = k/n) \right)$$
$$np = n\mu.$$

It follows that the mean of the sample proportion Y is $\mu = p$.

Let σ be the standard deviation of Y. As the variance of X is $np(1-p) = \sum(k-np)^2 P(X=k)$,

$$np(1-p) = \sum(k-np)^2 P(X=k)$$
$$np(1-p) = n^2\left(\sum(\frac{k}{n}-p)^2 P(X=k)\right)$$
$$np(1-p) = n^2\left(\sum(\frac{k}{n}-p)^2 P(Y=k/n)\right)$$
$$np(1-p) = n^2\sigma^2.$$

It follows that the variance for the sample proportion Y is $\sigma^2 = \frac{p(1-p)}{n}$, and therefore $\sigma = \sqrt{\frac{p(1-p)}{n}}$.

CPSIA information can be obtained at www.ICGtesting.com
Printed in the USA
BVOW01n1841110913

330800BV00014B/22/P